The Imagined World Made Real

HENRY PLOTKIN

The Imagined World Made Real

TOWARDS A NATURAL
SCIENCE OF CULTURE

ALLEN LANE
THE PENGUIN PRESS

ALLEN LANE
THE PENGUIN PRESS

Published by the Penguin Group
Penguin Books Ltd, 80 Strand, London WC2R ORL, England
Penguin Putnam Inc., 375 Hudson Street, New York, New York 10014, USA
Penguin Books Australia Ltd, 250 Camberwell Road, Camberwell, Victoria 3124, Australia
Penguin Books Canada Ltd, 10 Alcorn Avenue, Toronto, Ontario, Canada M4V 3B2
Penguin Books India (P) Ltd, 11 Community Centre, Panchsheel Park, New Delhi – 110 017, India
Penguin Books (NZ) Ltd, Cnr Rosedale and Airborne Roads, Albany, Auckland, New Zealand
Penguin Books (South Africa) (Pty) Ltd, 24 Sturdee Avenue, Rosebank 2196, South Africa

Penguin Books Ltd, Registered Offices: 80 Strand, London WC2R ORL, England
www.penguin.com

First published 2002
1

Set in 10.5/14 pt PostScript Linotype Sabon
Typeset by Rowland Phototypesetting Ltd, Bury St Edmunds, Suffolk
Printed and bound in Great Britain by Clays Ltd, St Ives plc

Contents

CONTENTS

Preface

This book was very nearly called *The Citadel Itself*, which was a phrase Charles Darwin used in one of his notebooks when considering the difficulties of scientific study of the human mind. The collective of human minds that we call culture is surely an even greater fortress waiting to be conquered by science. Hence the 'almost' title of a book which is a broadly based sketch of how to marry the social and the biological sciences. In the end a more direct title was chosen. Culture is precisely what the title indicates: 'the imagined world made real', and potently real at that.

Most social scientists will wince at what they will see as yet another attempt to invade their territory, but an eventual synthesis of the social and natural sciences is as inevitable as day following night. Science simply will not stand still. However, getting the synthesis right is going to take at least as much of an adjustment on the natural science side of things as it will on that of the social sciences, an adjustment which is going to step on the conceptual toes of some biologists. That is the reason why some two thirds of this book is biology – a figure that depends on acceptance of cognitive and comparative psychology as biology. If the social sciences are to be incorporated into biology, we had better get the biology right. And doing that will not be easy. One hundred and fifty years after Darwin the shape of evolutionary theory continues to change. Despite the recent resurgence of creationism under the banner of 'intelligent design', nobody in science doubts that evolution has occurred; that is, that life has been transformed in time, and all creatures derive from original living forms of around three-and-a-half billion years ago. Our own species does not stand outside this process. It is the 'how' of evolution that continues to be debated. If, as we grapple with the difficult problems that human culture presents, old, established concepts have to be modified, or even have to give way to newer ideas, well so be it. Biological purists might not like it but we are going to have to become more pluralist if something as complex as culture is to be understood within the framework of biology.

Why bother with culture at all, or, at any rate, why at this stage in our limited understanding? Why not stick with what Darwin was obliquely referring to as the citadel itself and move on to culture when we have a better understanding of the evolution of the individual human mind? The answer is simple. Our minds are powerfully formed by culture, which is simultaneously constituted of individual minds. In a very real sense you cannot do a science of mind without also doing a science of culture, and the reverse holds too. You cannot understand culture without psychology. And, of course, none of it can really, fully, completely, be understood away from the light cast by the central theorem of biology – evolution. If our minds have evolved in significant ways from the Miocene ape ancestral to ourselves and modern chimpanzees, which of course they have, the fact that we are creatures of culture has played no small part in that evolution. That is why this book weaves a path from social constructions, through the evolution of intelligence, on to theories of cultural evolution as a process, then to psychological mechanisms, the deep problem of the levels and units of evolution, and back to social constructions, before concluding with a brief survey of the links that must, and are, made between culture, psychology and evolution. It is a bit of a roller-coaster ride. A gentler metaphor is that writing the book felt at times like weaving a tapestry, with different threads drawn from the social and biological sciences. Some chapters are entirely drawn from the warp of biology, others from the weft of the social sciences. The first and last chapters run them together.

A word of explanation is needed about the phrase 'social construction'. Social constructions are things, like money and marriage, that exist only because people agree that they exist. They are entirely creations of human culture and do not exist outside of it. They are at the heart of culture and the most difficult part of culture to understand within a biological, or any other, framework. That is why they figure so prominently in this book. Only if you can claim to understand social constructions can you claim to understand culture. Within the context of the enormous complexity of social constructions, reef knots and using a spoon and fork are very small beer indeed. Culture is many things, this being one of the mantras of this book. There is no question, therefore, that motor skills like tying knots and using utensils are a part of culture. But they are trivial parts of culture, and explaining them by science in terms of how they are acquired tells us little about how we can explain someone's understanding of what an economic migrant is or what is patriotism.

Do not, however, confuse social constructions with social construc- tivism. The latter in its simplest and strongest form is the view that nothing escapes the forces of culture and that all knowledge, including science, and all values are relative to local conditions. This is the world view that tells us that the works of Shakespeare are no better than the scribblings of the local poet and that Newtonian mechanics is wholly a product of 17th-century English society. The ideology of social constructivism is a particular form of social construction. There is a sense in which science itself is a social construction because it is a product of culture. But it is a social construction with a difference. There are things out there, beyond our minds and cultures, of which science can and does gain knowledge. That the Earth circles the sun is a fact that can have no cultural spin. Sometimes local social events do influence the path of scientific understanding, an example of which is raised in chapter 2 of this book. But local influence is always eventually overcome by scientific progress. It is one of the greatest difficulties that a science of mind and culture must overcome. We must situate such science outside the influence of our own minds and cultures, and that is no easy thing to do.

The relationship between the social and natural sciences has always been uneasy at best and approaching open warfare at worst. This is because that relationship speaks to deep issues concerning the nature of humanity, how we understand ourselves, and how we conduct ourselves in the light of that understanding. No other area of science sits so uncomfortably close to ideology. The relationship is also fraught because social scientists often see biologists as intellectual hooligans in hobnailed boots who trample upon the subtlety and complexity of culture. I have sympathy with this view and have tried to tread as softly as possible. Critics might go so far as to accuse me of treading so softly that all I have done is tacked a lot of biology onto accepted social-science views of what it is about culture that matters. I would put it differently. I think culture is what social scientists have been studying for over a century and we ignore their views at our peril.

One cannot complete a book on humans and evolution just months after the publication of the results of the human genome project without making some reference to its findings and meaning. In fact its meaning, outside of medical application, is unclear. So too is the reduction by about two thirds of the estimate of the total number of human genes. It did not need the human genome project to tell us what has been known for some years, that we humans share many of our genes with other creatures, including those, such as fruit flies, from which we are separated by something like half a billion years of evolution. Nor does the total number of genes, between 30,000 and 40,000 as opposed to the previously thought figure of 100,000, alter much the massive disparity in numbers with the many hundreds of billions of nerve-cell connections in our brains. It is that disparity that remains the challenge, not the absolute numbers of genes. Not even the most unthinking of evolutionary psychologists has suggested that any of the connectivity of the brain is absolutely determined by genes. The construction of the brain through developmental processes is one of the great unsolved problems of science. We also know very little of the details of how the brain generates our psychological processes and mechanisms, some of which result in our being creatures of culture. However, in the last few decades there has been a significant growth of opinion within psychology that says the human mind is not a blank slate at birth, and that opinion is held by psychologists of one sort or another who have absolutely no interest at all in, and sometimes no time for, evolutionary psychology. Well, if the mind is not a blank slate at birth, something is writing on it prior to that, and this writing cannot all be accounted for by development. Development begins with a set of genes operating within a complex cellular environment that is also strongly affected by activity in neighbouring cells. Development is a very complicated matter. But that does not negate the link between the information in our genes and the evolutionary past of our species

– that is, after all, the only link we have with that past. If you believe that evolution has sculpted, even if only minimally, the structure and functions of our minds, then some of those 30,000 to 40,000 genes must have a part to play in that writing on the slates of our minds. We do not yet know how this occurs. But the evidence that it does is now overwhelming.

I have written this book for students and colleagues. I have, however, also written it for people at large. I find the task of writing science for non-scientists compelling. If you can't explain to your friend or neighbour why you think about a problem in the way that you do, you are not thinking very clearly. Writing for non-scientists, though, has consequences. One is that you cannot tell everything, because the detail becomes boring, if not confusing, to all but the specialists. I hope colleagues will forgive this. Another is that references to who did or said what where is dispensed with, which I do not regret. The academic style of 'Smith (1968) and Jaysaran (1982) have shown', with long reference lists at the back of the book, becomes 'Tom Smith and Vijay Jaysaran have shown', with short recommended readings for the interested at the end of each chapter.

There is a third consequence of writing for a general readership. This is that the book must be self-contained. Asking the reader to go and look up elsewhere what group selection is and why it is a controversial issue means losing that reader. For that reason I have included summaries of contemporary evolutionary theories and controversies, which are concentrated in the first and sixth chapters, though touched upon elsewhere. Those with prior knowledge can quickly skim through these.

The structure of the book is straightforward. Chapter 1 lays out the general problem of the relationship between the social and natural sciences, and an account of evolutionary theory as providing one of the conceptual toolkits for building bridges between the two. The unoriginal premise of this book is that it is in the understanding of the mind that the biological meets the social. An ideational approach to culture – that is, the notion that one way of viewing culture is in terms of shared knowledge and beliefs – is another basic theme of the book. Culture is seen as a particular expression of human cognition. For this reason chapter 2 considers the evolution of intelligence, with the specific aim of slaying two dragons: the genetic reductionism that so troubles social scientists is one, and the notion of the *tabula rasa* is the other. A general survey of anthropological schools of thought is given in chapter 3. Chapters 4 and 5 consider process and mechanism accounts of culture respectively. Because culture is, by definition, a property of groups of people, chapter 6 is taken up with the weighty issues of what the units of evolution are and how culture can be seen in this context. Finally, chapter 7 surveys what I think are the most important issues raised by social-science approaches to culture, and specifically their dealings with social constructions. It is in this last chapter that culture is slightly redefined in terms of the sharing of created, inventive knowledge and belief – that strange thing called social reality. Culture is imagination made real, and links between this conception of culture and some of the processes and mechanisms of previous chapters are explored. There is a case to be argued that one might best start with the final chapter, then go on to chapter 1, and end back with chapter 7. Well, I leave that to the reader.

I have spent a large part of my life reading and listening to what others have written and said on these matters. That is the lot of an academic. The names that appear in the following pages represent but a tiny fraction of all these influences. Others have been kind enough on occasion to listen to me and comment in turn. Too many to name individually, they all have my thanks for teaching me so much. I am especially grateful to Heléne Joffe and Lucy Yardley for guiding me into my readings in the social sciences.

Henry Plotkin
London, June 2001

I

Marrying the Biological
and Social Sciences

There was a time when humans were thought to be different from all other animals in a variety of ways. For thousands of years, despite living closely with domesticated species, humans were thought to be distinctive in having rational powers denied all other creatures. Five hundred years before the birth of Christ, the Greek philosopher Heraclitus declared there had been two forms of creation: Gods and men, with their rationality, and brutes, which are irrational. Non-human animals, it was held, behave impulsively, instinctively and without reflection. Learning, thought and reasoning were the exclusive domain of our species, it was later argued by Christian theologians such as St Thomas Aquinas, because these mental powers allowed us to tell good from evil, a judgement that determines our fate after death; indeed, they were an essential part of this process. Animals had no afterlife, so rationality of any kind was not necessary to their existence.

The first controlled demonstrations at the turn of the 20th century of learning in animals, by Pavlov in Russia using dogs, and Thorndike in the United States, whose subjects were cats, were the beginnings of a century of science that has gradually whittled away the differences between ourselves and other animals. In addition to relatively simple forms of associative learning, a variety of animal species, mostly mammals and birds, have been shown to be able to reason and to solve problems, to have quite stunning memory capacities, and to be able to acquire complex motor skills, demonstrate an elementary understanding of number and use tools. In recent decades, there have been claims, albeit much disputed claims, that chimpanzees, our closest living relatives, can learn human languages. It had begun to seem that the only distinctively human characteristic left was culture.

In a later chapter we will consider in some detail what exactly is meant by culture. For the moment let's settle on the idea that humans can gain information through three channels. One is the genes we inherit from our parents at conception. The second is our individual capacities for learning and intelligence without the intervention or mediation of others. The third is information gained from others. Culture is the consequence of the last of these. Skills, knowledge and beliefs are shared by members of a social group and comprise their common culture. And this is indeed distinctively human.

Or is it? It has been known since the 1960s that spatially separated populations of songbirds of the same species show variation in their regional dialects. Somerset chaffinches share variations in basic chaffinch song that are different from the variations in song of Kent chaffinches. In the United States, the white-crowned sparrow shows equally remarkable variation in song as one moves around the bay of San Francisco. It is known with empirical certainty that such variation is the result of young birds learning their local dialect by listening to the neighbourhood song of conspecific adult male songbirds. According to our simple definition of culture as information acquired from others, this looks like culture of some kind.

Different species of monkey have been reported as showing similarly circumscribed manual skills that they may have learned by observing others, though there are disputes (there are always disputes among scientists) as to whether these really are instances of cultural transmission of behaviour. What cannot be disputed is the accumulating evidence of some kind of culture in wild populations of chimpanzees. In 1999, nine primatologists reported in the journal *Nature* on seven long-term studies of different populations of chimpanzees, covering some 151 accumulated years of observations. What Andy Whiten – the first author, a primatologist based at St Andrews University – and his colleagues told us was that 39 separate behaviours, relating to tool usage, grooming, signalling and courtship, are widespread and customary in different populations, but different in how they vary from population to population. For example, two of the populations live less than 200 kilometres apart in Tanzania in near-identical ecologies and belong to the same subspecies of chimpanzee. Yet while food-pounding is customary among the Gombe chimpanzees, it is never observed in the Mahale group. Equally striking are behavioural differences between the separated West African populations of the Ivory Coast. One group cracks open nuts by placing them on a rock 'anvil' and striking them with a stone or wooden club; such innovation has never been observed east of the Sassandra-N'Zo river, even though there are nuts there aplenty. Differences in ecology and genetics cannot explain such behavioural variations confined to spatially isolated groups. These animals are learning from each other.

But is this culture? Well, yes, it is according to our simple definition of culture as learning from others; and we will return to the issue of animal culture – call it protoculture if you prefer – at several later points in this book. However, it simply does not compare with what I, the writer, and you, the reader, are doing right now as an exercise in sharing knowledge and beliefs. As someone once remarked, if animals had culture of the kind we humans have, the ducks would be holding seminars on imprinting (following the first moving object seen, which is what ducklings do) and the rats giving lectures on the Skinner box (an apparatus much used in laboratory experiments on learning in rodents). Of course, these animals don't hold seminars and give lectures, so they do not have culture of a quality anything like that possessed by us humans. Nonetheless, the lesson to be learned from songbirds and chimpanzees is that culture is not some single thing. It comes in different forms. Nor does any one view, approach or method of study have a monopoly on the understanding of culture. So what we will be looking at in the pages to come is a number of questions, and perhaps some answers too. But in addition to having to ask more than one question, what I want to do is ask the very hardest question of all.

Human culture is the most complex phenomenon on Earth, comprising as it does the collective skills, knowledge and beliefs in the minds of people that make up a culture. The question at the heart of this book is can human culture be brought within the explanatory framework of the natural sciences? And while culture itself is not just one thing, at the heart of human culture are social constructions, which are also extraordinarily complicated and strange phenomena. So the hardest question that can be asked is whether social constructions specifically can be understood by the natural sciences. If we find the answer is yes, or can at least begin to frame the answer in a way that suggests the question can be answered in the affirmative, we will be making progress towards solving some of the big unanswered questions in science.

Bearing in mind, then, that culture is many things and that there are many approaches to understanding it, what I want to do in this chapter is to begin to pose some of the basic issues in terms of the hardest questions. What is a social construction? Why is it important to bring the natural sciences to bear on them? And which of a number of possible approaches should be used? Later chapters will serve to broaden the way we look at culture at large. So, let's begin with an example of a social construction.

In 1948 a 'white' Nationalist government came to power in South Africa. Over the next few years, by political manoeuvre and sleight of hand, the constitution of the country was changed. As a result, people who were not of European descent – that is, non-white people – were systematically deprived of the few political rights they had had, were frequently uprooted from their homes and homelands, and were subject to Draconian laws relating to restrictions on work, movement and basic human rights such as living with whom they chose. These political acts and the laws they gave rise to were known as apartheid, a hybrid English–Afrikaans word that means separateness. Apartheid was meant to separate white people from all other South Africans to the point that what contact did occur would be entirely to the advantage of the whites. The ultimate intention of the white government was that a small minority of the population of South Africa would live almost exclusively in much the greater area of land, including of course those parts that were rich in mineral resources and agricultural produce, with the majority of its citizens squeezed into small and unproductive regions strategically placed to provide cheap labour for white enterprises.

In the end, and for many reasons, apartheid failed. The exclusion of non-white peoples from the political process was abandoned in 1990 and the first elections based on universal suffrage took place in 1994. However, for almost four decades apartheid was not just some preposterous and ludicrous notion, some dream for future action, but a politico-legal system enforced by coercion, infringements being punished by fines, imprisonment and worse. And underpinning apartheid was baaskap. *Baaskap* does not have a direct equivalent in English. Literally translated it means 'bossedness', but it also carries the implication that the natural role of white people is that of superiors, the natural leaders and bosses of people who are not white. The assumption of white superiority dates back to the start of the colonization of Africa and other parts of the world by Europeans in the 16th and 17th centuries. It was present all the way through the 18th and 19th centuries. The 'scramble for Africa' in the 19th century, by which European powers came to rule over virtually the entire continent, was fuelled in part by a belief in the superior abilities and status of Europeans. Few colonists did not subscribe to the racist belief that white people were superior to non-white people. So, from the start of European colonization of Africa, the indigenous peoples of Africa were considered by the colonialists and white settlers to be at best second-class citizens, whether rule was informal, as in the more remote regions, or formalized, as in the more settled areas. Apartheid was the endpoint of this belief and its formal translation into an oppressive political system.

Apartheid and baaskap have a number of features which should be noted. First, baaskap was, and perhaps still is to some extent, an idea held collectively by almost all members of the culture of white South Africa. Through upbringing and education (in its broadest sense), most white South Africans agreed with the belief in white superiority. Baaskap was a powerful social force that existed because the people within a particular social group agreed that it existed. I follow the philosopher John Searle in holding up *something which exists because of collective agreement* as the defining feature of a social construction. Baaskap was, and remains in many parts of the world, a social construction.

Second, the causal force of social constructions should not be underestimated. If causal force is measured by the power to affect people's lives, social constructions are very significant entities. Baaskap has affected the lives of millions of South Africans over hundreds of years in every detail, including in some cases, in the ultimate expression of causal force, bringing about their end. Baaskap is not, in this respect, an unusual social construction. During the period of recorded human history most wars have been caused by social constructions such as religion, ideology, national honour, patriotism and control of resources such as money markets. In these wars people have been killed in huge numbers. That is some causal force.

Third, collective agreement always finds expression in material forms. Apartheid, as a political system that gave rise to a series of laws, found material expression in passbooks (documents non-white people had to carry that gave them permission to be where they were at certain times), the distribution of people in space and the distribution of wealth, among many other things. The material expressions of social constructions feed back in important ways to the maintenance of the collective belief. Apartheid deprived most South Africans of all but the most rudimentary education and denied them jobs in all but the most menial positions. This meant most South Africans were ill equipped to deal with the world of the 20th century. You do not have to be stupid to fail in the world if you have not been taught to read. So most non-white South Africans appeared incompetent and poor because the physical forces of apartheid *made* them incompetent and poor. And that fed back to reinforce the belief in baaskap cherished by those who were advantaged by apartheid. There is a complex dynamic between social constructions and material structures of the world, the latter being caused by, and in turn affecting, the former.

Fourth, social constructions are always a product of history. It is clear from the brief account given here that baaskap did not spring fully formed into South Africa with the 1948 elections. Racist social constructions by Europeans about non-Europeans are at least hundreds of years old. It is also possible that the social construction of baaskap is just one of a family of social constructions concerning 'other' peoples (or outgroups, to use another terminology) that occur within all cultures. That is, maybe social constructions about people who are different, perhaps by simply not being members of a particular culture, by being strangers, are a universal of all human cultures. The possibility that some social constructions are cultural universals will be considered in chapter 7. If this conjecture is correct, the provenance of baaskap is very ancient indeed.

Fifth, social constructions are central and essential features of contemporary human cultures. They are certainly not the only features of culture, but no study of contemporary culture would be complete without reference to the cluster of social constructions that lie at its heart. Social constructions do not, of course, simply involve racist beliefs. Justice, money, money markets and patriotism are examples of other forms of social construction. Baaskap and apartheid are simply illustrative of the general point just made, that social constructions are essential features of culture. Prior to the 1990s, white South African culture was steeped in the notion of white superiority, and baaskap adversely affected the lives of all non-white South Africans and their culture. No social-science understanding of South African culture was possible that did not take into account the social construction of baaskap.

Culture, social constructions and natural science

Would it still be necessary to consider the social construction of baaskap, along with other factors, if one wanted to expand the enquiry and try to explain apartheid South Africa within a natural-science framework? The answer, surely, is yes. The framework would not matter. Baaskap was at the heart of South African culture for a long time and that is what would have to be explained, whatever science were being used. It follows from this that if the natural sciences, specifically biology, are ever to get a conceptual hold on culture in general, and any specific culture in particular, then it had better come to explanatory grips with social constructions.

Why, though, should we seek for a natural-science account of human culture? This is a deceptively simple question, to which there is no easy answer, raising as it does a very big issue. So I will give several answers. The first, and shortest, is that the natural sciences include the social or human sciences. Humans do not stand outside nature. We are as much a product of evolutionary forces as are any other animals on the planet. Despite human intelligence, as will be argued in the next chapter, we are also as much a product of our genetic endowment, realized in nervous systems and neural-network structures through development within particular environments, as are any other intelligent animals with nervous systems. Of course, humans are also different from other species, including their closest living relatives, and those differences define the special status of the social sciences. But these differences do not remove the study of humans from the sphere of the natural sciences.

The second, slightly longer, answer is that the natural sciences are one of the great achievements of our species, especially of the 20th century. The structure of matter and the origins and history of the universe are understood now as never before, this understanding being enshrined in the theories of relativity and quantum mechanics. While the first part of the century was dominated by the physical sciences, the second half saw an explosion of knowledge in the biological sciences, starting with the famous Watson and Crick paper of 1953 describing the molecular structure of DNA (deoxyribonucleic acid), the chemical whose linear structure codes the information by which genes, via the production of proteins, generate the structure and function of living creatures. As in the physical sciences, knowledge of the very small, of the structure of cells down to the molecular level, has been accompanied by an increasing understanding of the very large, namely the interactions between populations of organisms in the study of ecology. Landings on the moon, planetary probes and cloning bear witness to equally spectacular achievements of applying 20th-century science. So, if you want to make a claim to knowing and understanding something, then the strongest, the most convincing, case must be couched within the province of natural-science explanation.

The reader should be clear as to what is meant by *explanation* and *knowledge* in this context. Science is a hard taskmaster when it comes to knowledge. It demands the ability to make exacting predictions about phenomena that can be measured. This requires the uncovering of regularities within these phenomena, which must be described in laws and explained by theories that, above all, are cast in terms of causes. The result is often an account of the world that is counter-intuitive and far removed from ordinary experience. Hence the strangeness to non-physicists of the theory of relativity, the sense for many people that it is impossible for the forces of evolution to have brought about the astonishing and beautiful forms of the living world, and the apparent absurdity to most of the notion that the continents are ever so slowly floating about on the surface of the planet. Science need not, and often does not, trade in ordinary, everyday experience and explanation. The very unexpectedness of scientific knowledge attests to its unique power as a way of knowing and understanding the world.

This kind of disciplined knowledge is a long way from the understanding that, say, crops are to be planted at a certain time of year in order to gain maximum yield when that understanding is based upon mere trial-and-error practice rather than explanation. It is only when the mechanisms of plant growth, and the effects of light and temperature on those mechanisms, are known that we have a scientific explanation of crop cycles. The same applies to the human sciences. Here we are looking for something quite different from the folk psychology that allows us the better to get on with one another by using social skills appropriate to particular situations. We are questing far beyond the commonplace predictions we make about one another's behaviour. And while all science begins with description, the making of lists describing the phenomena we wish to understand, no matter how detailed, is not enough in any science, including the human sciences. Processes and mechanisms, often not directly observable, must be invoked to account for the phenomena we see, and those processes and mechanisms must be dressed up as causes. Hypothesized causal mechanisms lead to predictions of the 'What will happen if?' variety, and such predictions are then empirically tested, the results of those tests being used further to shape and sharpen our causal explanations. And so the process of science rolls on through the continual interplay of ideas and theories and their empirical testing by observation and experiment.

It is one thing to say that, in this respect, the human sciences must be no different from other science, but how does this picture of deep, non-intuitive knowledge in the natural sciences really play when it is applied to the human or social sciences? Has our understanding of human nature and human interactions kept pace with advances in the natural sciences? The answer is no, though there have been modest gains. A great deal more is known at the start of the 21st century about our sensory systems than was the case a hundred years ago. We know more too, if only a little more, about how we interpret sensory data – perception – and how we come to understand the world – cognition. There is also increasing understanding of how we put coordinated actions together to form complex behaviours. Rather lesser gains have been made in areas such as emotion and intelligence, and we are little further on in our knowledge of matters such as creativity and aesthetics than we were at the start of the 20th century.

On the other hand, the case for the advances made in the social sciences should not be understated. We really do know more now than was the case just a few decades ago about memory and attention, how we come to use language, how children acquire arithmetical skills, and much else. As the 21st century unfolds, the science of mind – psychology – will doubtless gain in strength and depth of understanding. Neuroscience, the study of the structure and function of the brain and peripheral nervous system, is growing in leaps and bounds. Stronger links will be established between the wet matter of the brain and the psychological mechanisms that drive mental life and behaviour. There is also no doubt that neurogenetics and behavioural genetics will strengthen understanding of how genes enter as part-causes into the way our brains and minds function. This is not scientific triumphalism. Anything seems possible in an age when serious attempts are being made to clone tissues as replacements for diseased or worn-out body parts, and to make connections between computer chips and neural networks in living organisms – attempts that will certainly succeed in the near future given the success of cochlear (part of the inner ear) implants connected to auditory nerves in restoring hearing to thousands of deaf people. Even the elimination of death becomes a possibility. A preposterous idea, perhaps, but it provides us with an important thought experiment because it tells us about the power of culture, which is what science is, over our material selves. In any event, with such prodigious achievements, a sound scientific understanding over the next few decades of human cognition seems eminently achievable, if not small beer.

Yet the natural sciences have made no impact at all, or at any rate very little, on the attribute that seems, above all, to make us human. This is culture, and the human capacity to enter into culture, which is an awesomely complex phenomenon. The natural sciences so far have contributed little to our understanding of culture, and nothing at all to an understanding of social constructions. But that doesn't mean this will always be the case. It simply must be that an understanding of human culture will eventually be reached under the banner of the natural sciences. But this will only occur through the conceptual marriage of the biological and social sciences, which is one of the great remaining syntheses to be brought about in science, and which will indeed happen over the course of the next hundred years or so. This book is an exploration of some of the ways in which this might happen.

There is a third and even longer answer to the question as to why a natural-science account of culture in general, and social constructions in particular, is something worth pursuing, and why we should not rest content with current social-science accounts of culture and social constructions, no matter what judgement is made of their excellence as social science. In part, the answer is one of personal taste and personal history. But I also believe there are deeper issues at stake here, delicate and complex matters involving some very old disputes and deep divisions over the nature of human beings and how best we should study ourselves. Being so long, the answer is best given in two parts.

The first involves the nature of the social sciences themselves. It is no simple thing to distinguish between the social and the natural sciences, and to say 'here the one ends and the other begins'. And there are, of course, a number of distinct disciplines within the social sciences, each with their own differences and boundary lines (and boundary disputes) between themselves and with the natural sciences. Yet a broad distinction has been drawn, and continues to be drawn, between the natural and social sciences which has its origins among German thinkers of the 19th century, notably Wilhelm Dilthey. There is a difference, they maintained, between the natural sciences, *Naturwissenschaften*, and the sciences of the mind, *Geisteswissenschaften*. While the former are exact sciences based on experimentation, quantification and precisely stated laws and theories, the latter are closely allied with the other human sciences of anthropology, sociology and economics (even though these emerged as separate disciplines only after the distinction had been drawn), together with philosophy, history and the rest of the humanities. Dilthey's pithy comment that 'We explain nature but we understand human beings' is a clear and succinct summation of the distinction and the supposed differences in method that each required.

For many psychologists, as well as other kinds of social scientist, *verstehen* became the preferred approach. *Verstehen* involves interpretation, hermeneutics and empathy, not experimentation and quantification. It aims to reconstruct meaning from the individual's point of view; that is, the perspective is internalist and relative, meaning being inextricably linked to each individual's own context. This is to be contrasted with the natural sciences, which aim at explanation within an externalist perspective of general laws of the world and the analysis of the causes of events in that world.

The subsequently drawn distinction between the nomothetic (science-based general laws) and the idiographic (the uniqueness of the individual) seemed to complement Dilthey's distinction. That the uniqueness of the individual is the result of universal processes and mechanisms that can be understood within the framework of natural science is very much a 20th-century world view, and even now, at the start of the 21st century, is still not universally accepted. Certainly the majority view of the 19th century was that prediction, experimentation, measurement and causal explanation were fine in the natural sciences but inappropriate for the social sciences. The complexity of the human mind, of consciousness and of social interaction is so great that measurement is less relevant, if possible at all, than non-explanatory understanding, in which the individual is understood as an individual whose own understanding is uniquely determined by their particular circumstances. There is a direct line from Dilthey to the radical relativism of postmodernist social psychology in particular and the social sciences in general.

It is worth noting that the prevalence of the distinction between the social and natural sciences is attested to by none other than Wilhelm Wundt himself. Wundt was one of the principal founding fathers of experimental psychology and the first director of the first experimental psychology laboratory, established in Leipzig in the 1870s. Wundt held the view that social psychology (*Volkerpsychologie*) was a distinctly different enterprise from experimental psychology, the domain of which included such subjects as sensory processing and memory. But the social scientists for whom culture has been most central are anthropologists and sociologists.

One of the major issues among, and divisions between, such scholars has been the position taken on the relationship between individuals and the social groups of which they are a part. Methodological individualists (the jargon goes with the area) asserted the primacy of the individual over the social group; that is, the latter was reducible to the individuals of which it was made. Methodological holists had a quite opposite view with their counter-claim of the irreducible and autonomous nature of social groups, the properties of which were held to be different from those of individual people. A company or state is an entirely different entity from the employees or citizens within it. The image that is conjured up is of a 'superorganism', a notion and a phrase invented in the 19th century by Herbert Spencer in England and Emile Durkheim in France, and propagated into the 20th century by the founding fathers of American anthropology, Franz Boas and Alfred Kroeber. There is a mountain of scholarly writing about the superorganism, how it emerges from social interaction, the necessary conditions for such emergence, and, indeed, if it can be said to exist at all or whether it is merely a metaphor or some other kind of helpful concept.

Unsurprisingly, there are also social scientists who take an intermediate position between the extremes of methodological individualism and holism. One example well worth following up is the British social anthropologist Tim Ingold, who adopts what he calls a 'constitutive' position. He rejects the superorganism-with-a-mind-of-its-own notion, but insists on the crucial role in human development of the process of social constitution of one another through a complex process of mutual involvement. Value, purpose and belief are not properties of a superorganism of which we are all a part, but nor is their locus the individual. Rather they are constituted by relationships with others. The social world is a source of values, purposes and beliefs, not the vessel into which they are poured. Put another way, these things are social products manifested within a social context, rather than formed by people independently of, or outside, social life and then brought to social interactions.

Such an approach, which will be expanded on in chapter 7, does not eschew the possibility of making connections with the natural sciences, either by way of evolutionary theory or by way of neuroscience. While happy to castigate what is seen as poor or primitive attempts to make these links, by and large those who follow this approach seem content to leave this job to others. In the last century some of those making these connections have been other kinds of social scientists, such as social psychologists, personality theorists and clinicians, whose approaches have been empirical, even experimental, and who have reached out to the natural sciences via cognitive science, game theory and brain function. By definition, though, these are physicalists or material social scientists, i.e. natural scientists of the mind. It has usually been the case, however, that they have not been concerned with culture in general or social constructions in particular.

This thumbnail sketch shows a wide array of possible approaches to the social sciences. How, though, is one to make sense of them? Which is the right way and which are in error? Let's return to Dilthey with his original distinction between 'explaining nature' and 'understanding human beings'. Implicit in this is the claim that human beings, and their study by social science, are inherently different from natural phenomena and their study by the natural sciences. This is not necessarily to claim that humans are 'unnatural', for some reason lying outside nature's realm, though some have espoused such a view. It certainly *does* involve the claim that somehow, for some reason usually entailing certain human attributes such as consciousness or intentionality, humans are different from all other things and cannot be studied or explained by science in the way that stars, rocks or even (perhaps) chimpanzees can. In essence, it is a denial of the unity of science. And this brings us to the second part of the longest answer to the question of why a natural-science account of culture is worth pursuing. The unity of science is a beautiful idea, and beautiful ideas should be pursued.

The possibility of a unified science has a long history that stretches back at least two thousand years. It comes in different forms. Some demand a unified language, some a unified methodology, others a unified process, and to some, perhaps the majority of those who seek a unified science, the aim is intertheoretic reduction. Intertheoretic reduction, in ideal form, aims for one basic theoretical level, normally supposed to be some theory of physics, to which all other theories can be reduced. The precise nature of the conditions of reduction are complicated and stringent. Suffice it for our purpose to say that theory reduction occurs when one theory is related to another in such a way that the first is able to be subsumed under the explanatory mantle of the second. By a stepwise process, the social sciences will be reduced to psychology, psychology will be reduced to neuroscience, neuroscience to biology, and biology to chemistry and physics.

Now, many philosophers of science, who are the people who set the exacting requirements for intertheoretic reduction, consider these criteria can never be met, at least not in their totality and not across all levels of science. Furthermore, reduction might fail for reasons other than the formal and technical business of intertheoretic reduction. For example, in chapter 2 I will present an argument as to why the behaviour of intelligent animals, including humans, cannot be reduced to genetics. So perhaps the ideal of complete intertheoretic reduction will never be achieved. As an ideal, though, it remains strong and, incidentally, is feared by many as threatening essentially to sideline those sciences that are reduced to others. This is not the case, as will be argued in the next chapter. But to return to the main point, the possibility of absolute or even just partial reduction is not, in my view, the only criterion for maintaining a belief in the ultimate possibility of a unified science. A belief in the *possibility* of *some* reduction between *some* levels, wedded to an absolute requirement for a commitment to the discovery of explanatory causal mechanisms that account for regularities in the world, will do, especially if some of those mechanisms extend across levels. Thus the commitment is not to a complete and absolute unification via a complex chain of theory reduction to a small number of causal mechanisms. Rather it is to a looser view that connections between the sciences can and must be made; to the materialist view that there is no mysterious mind or culture 'stuff'; to a position that maintains that life, including mental and cultural life, is no more than chemistry and physics; and to the view that the world is knowable through science, and that science aims to provide causal explanations for everything, including mind and culture.

That is why, to come back to the original question, a natural science of culture is worth pursuing. It is all about the grand vision of a unified science. Dilthey's position denied this possibility. The relativism of postmodern analysis also denies it. As stated earlier, it is partly a matter of taste. Well, my taste is not for partial, fragmented and unconnected knowledge of the world, but for a unified knowledge based on causal mechanisms. If you are going to think, then think big. The very idea of a unified science is magnificent. It is one of the great goals of human thought.

The position of many social scientists, like that of Ingold, is in my view entirely compatible with the ideal of unified knowledge gained through a commitment to the possibility of a unified science. This does not mean that social science and culture are simple matters. To repeat, culture is awesomely complex. But it must be – it simply must be – within the scope of understanding of the natural sciences.

Possible frameworks

If we are seeking to marry the social and biological sciences using culture, specifically social constructions, as the basis for the union, we need to consider which area of biology is best suited to the enterprise. There are a number of possible candidates, and were this book being written 50 years from now, I do not doubt serious choices would have to be made to keep it to a reasonable length. But right now choices are limited by what we know, and for the present too little is known about brain structures and brain function as they relate to the mind to be able to link with a reasonable degree of certainty any area of social science, aside from language, to any of the neurosciences. To be sure, understanding, especially through functional imaging of the brain in normal conscious people employed in different mental tasks, is increasing apace. But we cannot yet make firm enough connections between our physiology and our culture. And while estimates are made that as much as half the human genome (the genome is the totality of genes, thought to be in the region of 35,000, that carry the information by which development results in each individual human) enters into the construction of our brains, firm and unequivocal evidence is lacking that links genes at particular chromosomal sites either to specific brain structures or to psychological mechanisms. There are tantalizing glimpses of a genetic basis for attributes such as language and intelligence, for some forms of psychological abnormality, and even for sex differences in social skills, some of which will be discussed where appropriate. But an account of any aspect of culture in genetic terms remains science for the future.

While we do not yet know very much about the genetics and neurology of culture, enough is becoming known about some psychological mechanisms to make them strong candidates for laying the foundations for the wedding of the social and biological sciences. Precisely which such mechanisms should be looked to, and quite how they bridge the conceptual gap between the biological and social sciences, will be taken up in chapter 5. Suffice it to say here that this is certainly one of the ways to marry the social and biological sciences.

There is another, overarching, framework that forms the conceptual background of this whole book. This is evolution and the theory of evolution, both unequivocally part of natural science. There are several reasons for using evolutionary theory as the principal bridge between biology and the social sciences. One has already been mentioned, and it is the most important. This is that our minds and social lives do not fall outside nature where they must be explained (or, rather, understood) by invoking non-material causes. There are no non-material causes. Biologists make the assumption that most forms and functions of living creatures, be they bacteria or buffalo, are the consequences of evolution – and what is not attributed directly to evolution is usually indirectly linked and always rooted in material cause. The social sciences study a coherent aggregate of human characteristics, and humans are animals, the products of evolution like all other animals. Therefore, that aggregate of human characteristics studied by the social sciences must also be products, directly or indirectly, of evolution. Special we may be, and indeed we certainly are. Culture is largely a human-specific characteristic and, chimpanzees and songbirds notwithstanding, there is nothing remotely like it anywhere else in the animal kingdom. But we are not so special that we fall outside natural law.

It is often said that 'so much of what we do, like wearing clothes, heating our homes, driving cars and flying in aeroplanes, is unnatural'. This is simply false. So much of what we do is *different* from the activities of other animals, and what we do now is *different* from what our ancestors did. The difference lies in the cumulative effects of culture. But our culture is a part of human nature, and human nature is a part of wider nature. To alter slightly the words of Stuart Kauffman, the American complexity theorist, 'We are at home in the world'; indeed, an advocate of a unified science would argue, as does Kauffman, 'We are at home in the universe.' I am inclined to take an even harder line. Wearing clothes and visiting a supermarket are not unnatural, just particular facets of being human. This also applies to altering the genetics of other species and to other advanced scientific methods such as cloning. Our behaviour only becomes unnatural when it decouples us from the ultimate selection filter of altered survival as a consequence of that behaviour. Put baldly, when culture conquers death, then, and only then, is it unnatural.

Despite a studied avoidance of any discussion of humans in his first book, it was Charles Darwin himself who, in subsequent works, began seriously to consider the consequences of applying his theory to our own species. There has been a great surge of interest in the last two or three decades in carrying this project further. It is, I suggest, one of the most profound movements in the history of attempts by humans to understand themselves. At its heart is the notion that we are as natural as any other creature in nature.

The second reason for using evolutionary theory as the marriage broker (to maintain the metaphor of this chapter's title) is that it is the only way we have of explaining functional adaptive design in living things. I will *not* adopt the view that human culture, and our capacity to enter into it, is a single attribute with an adaptive design that has evolved to carry out some specific function. Culture is not like our vascular or visual systems. I will be arguing, however, that culture is the outcome of a number of separate features of our minds, each one of which is a separate organ of mind which did evolve specific adaptive features with specific functions. Culture is the happy outcome of the conjunction of these separate organs. And, as with the slashing teeth of predators and the fleetness of foot of their prey, such design can only be explained through evolutionary theory.

A third reason for the dominating presence of evolution in this book is architecture. A conceptual marriage is rather like bridge-building, the bridge spanning the gap between two separated points in conceptual space. It is a long way from cellular biology to social constructions, and the only theory in the natural sciences that can span that distance is evolution. Different architects will, of course, build bridges of slightly different design. Recently, for example, two eminent biologists, Eörs Szathmary and John Maynard Smith, have delineated eight major transitions in the history of life on Earth. They begin with the transition of replicating molecules, the foundations of life, to populations of such molecules in some kind of compartmentalized structure. Other transitions include the evolution of the genetic code, sexual reproduction and multicellularity. The final transition is from primate to human societies with the appearance of language. Well, this is a good example of thinking big. It's a huge bridge that they build and its structure is evolutionary. My bridge is of altogether more modest dimensions, but, as for Szathmary and Maynard Smith, evolution is my basic conceptual material.

For all these reasons, the remainder of this chapter will be concerned with evolutionary theory and its application to our own genus and species, *Homo sapiens*.

Evolution and the theory of evolution

The philosopher of mind Daniel Dennett thinks the theory of evolution is the greatest achievement of science. Richard Dawkins, the British biologist, has suggested that one way of determining how advanced any form of extraterrestrial life is, when finally we make contact, will be to ascertain whether the aliens have discovered the theory of evolution. A president of the United States, while on the hustings some years ago, responded to a question by noting, with something of a sneer, that evolution is just a theory. Twice in the last century, evolutionary theory has been at the centre of sensational legal trials over the issue of whether it should be taught in the public-school system of, of all places, the United States, where something in the region of 80–90 per cent of world science is carried out. The state of Kansas in 1999 withdrew evolutionary theory from its school curriculum (along with all reference to the Big Bang). And I am the fond possessor of a remarkably confused diagram that appeared in a 1983 issue of the *Bible Science Newsletter*, laying all the ills of the modern world – ills, at least, as this publication sees them – including euthanasia, runaway taxation (truly) and the acceptance of homo-sexuality, at the door of evolutionary theory. What is one to conclude about a theory that raises such extreme and opposite reactions, that is hailed as one of humankind's most profound accomplishments, dismissed as just another creation myth, or reviled as a truly evil idea? Well, given that creation myths are powerful narratives in all cultures, one surely has to conclude that, right or wrong, for good or ill, evolution is an extremely powerful idea, the creation myth of modern times. The odium in which it is held in some quarters derives, of course, from evolutionary theory's claims about human origins. The great strength of the theory is its explanatory power. As we will see, it is, in fact, no myth.

Given the emotive associations of popular accounts of evolutionary theory with highly coloured, and often inaccurate, phrases such as 'survival of the fittest', 'nature red in tooth and claw' and 'the naked ape' (us humans, that is), a safe place from which to start a brief account of evolutionary theory is in the contrast between order and flux in the world. In general terms, the view from my study window is pretty well identical to that from any other window. I see *diversity* in the form of different kinds of trees, shrubs and birds (and, if I had microscopic vision, as I peered down into my carpet and the earth of the flower-beds in the gardens I would see diversity several orders of magnitude greater than I can with the naked eye). I also see *change* in that clouds are being driven across the sky, branches are moving in the wind and the berries on the rowan tree are being eaten by blackbirds that are constantly shifting position. That is the change I see with a quick glance. Measured in hours rather than seconds, change would extend to temperature, light levels, humidity, and water levels in the soil that contains such an astonishing range of living forms. Measuring over an even longer time, sampling as widely as possible and using sufficiently sensitive measuring devices, one would observe that nothing is unchanging. Yet I also see *order*. There is a coherence in what I see out of my window. The dozen or so trees that are visible are discrete entities separated from one another and from the birds that fly between, and perch on, them. Those birds are bounded by their skin and feathers, and while I know that birds, like ourselves, contain whole ecosystems of bacteria, were they visible those bacteria would themselves be revealed as bounded entities. Furthermore, the spatial positions and paths of trees, birds and bacteria are relatively constant. I know where the blackbirds are nesting, just as the blackbirds know the position of the rowan and cherry trees on which they feed. Because my study is in London, I also see an imposed order of Victorian houses, fences and a road that has endured as a fixed order for over 150 years. Some of that order, then, is created by ourselves. But we humans are not alone in imposing order on the world, even if we may be excessive in the extent to which we do so. Many species of animal build nests, excavate burrows and create other structures, sometimes of astonishingly complex order. And the order I see from my window will be maintained for different periods of time. The roses will be gone in days, the berries and cherries in weeks, the leaves on the trees in months and the birds in years. Yet every instance of order has the quality of enduring for a finite, sometimes quite long, time.

Diversity, change and order are explained by physicists and chemists as the consequence of living things being dissipative systems, to use the language of the Nobel prize winning chemist Ilya Prigogine. The dissipation Prigogine was referring to was of energy and matter in open, non-equilibrium thermodynamic systems. In closed systems, where energy is not exchanged with another system, order is minimal and entropy reigns. This is the second law of thermodynamics, which describes the drive of closed systems to states of enduring disorder. Our planet, however, as is every living organism on it, is an open system. Energy from the sun washes across the Earth. Some of that energy is captured by living things and used to drive complicated chemical machinery, creating gradients of negative entropy in a universe of positive entropy. The capacity of life to capture energy and use it to maintain its own highly unlikely and relatively enduring order, to run counter to the second law, is the reason for the diversity, order and change that I see from my study window.

What I have just described, ever so briefly, is a late-20th-century account, in chemistry and physics, of the state of the world. This was not an account available to biologists and other scientists 150 years ago. Yet, when they looked out of their windows, they too were struck by the same seemingly contradictory characteristics of the world. It was changing and diverse, yet ordered. How could this be explained, especially against the background Victorian religious view of a world created just a few thousand years before, in the short history of which different causes operated at different times and when change came it was, like the Great Flood, catastrophic?

One form of explanation was to challenge the assertion that Earth was created just four thousand years before Christ lived, and to assert a constancy in the causes that govern events on it. This is exactly what some geologists and philosophers were doing in the 18th and 19th centuries. In 1795, the Scottish geologist James Hutton proposed a 'uniformitarianism' by which the processes of geological change are constant, hence those observable today are the same as those that have always operated. The world is in a steady state as far as causes are concerned. 'No vestige of a beginning, no prospect of an end,' wrote Hutton, which was heady stuff for that time. Estimates of the age of Earth increased steadily as the 19th century progressed, ignoring the error made by Kelvin that for a time threatened the theory of evolution by so markedly reducing the estimated age of the planet. But an error it was. Hundreds of thousands of years became millions, and millions became tens and then hundreds of millions.

Another trend in findings that paralleled that of an older and older Earth was the discovery by biologists of ever more forms of life. Diversity and order were becoming increasingly obvious. But so, too, was change. Fossil-hunting became a positive rage in the 19th century, and the more the fossil record was probed, the more it became clear that fossil forms were evidence of change – as if species could somehow be transformed. And the older the fossils, the less they resembled contemporary life forms. If all of life had been created just once, and recently at that, how could living forms have become extinct while the diversity of the biosphere had increased?

None of this made sense unless the Earth was much, much older than was claimed by the Christian creation myth, and unless species could be transformed into different species. We now know that Earth is in the region of four-and-a-half thousand million (4.5 billion) years old. This is a difficult idea for the human mind that thinks Westminster Abbey is old at a thousand years. Grasping the magnitude of deep geological time is an important part of stepping into the conceptual shoes of evolutionists. And the possibility of the non-fixity of species, a startling idea even to scientists because it violated the axiom of Platonic essentialism – the notion of discrete and always separable forms, which had dominated thinking for over two millennia – was voiced by a number of 18th-century writers. The possibility of evolution was 'in the air' as the 18th century turned into the 19th, and the first theory of evolution was put forward by the French naturalist Jean Baptiste de Lamarck in 1809. In fact his lecture notes, according to Ernst Mayr, indicate a conversion to evolution and the emergence of a theory as early as 1800, but the lag in publication is entirely normal. Charles Darwin developed his theory in the 1830s and was only forced into publication 20 years later when another biologist, Alfred Russel Wallace, wrote to Darwin outlining an identical theory.

Lamarck's theory was wrong, while the Darwin–Wallace theory was remarkably close to evolutionary theory's current form. Despite Wallace's equal claim to the theory, Darwin's priority in terms of when he formulated it, as well as the enormous mass of indirect evidence and argument he had accumulated (not to mention his having powerful friends in the world of English science, which Wallace lacked), have resulted in its current form being usually referred to as NeoDarwinism.

Many biologists believe the essence of Darwin's thesis remains intact. NeoDarwinism extends the original theory by explicating many of the mechanisms and filling in much detail. At the heart of the theory were (and remain) three processes: variation, selection and transmission. Darwin was a magnificent observer. He did not just see the immediate picture before him, but was able to assemble a pattern from his observations as well as from the ideas available to him at the time. His travels around the world had revealed to him not just diversity but systematic variation. Perhaps the most famous example is the dozen or so closely related species of finch on the Galapagos Islands that now bear his name. Darwin believed they must all have originated from some single ancestral species, but how had this occurred?

The key for Darwin, famously, was his reading of Malthus, who, on the basis of assumptions about potentially unchecked population growth, had inferred a struggle for existence between individuals. In Darwin's time, however, it was believed that populations of organisms maintained relative constancy of size because of limitations of resources, and thus Darwin was 'well prepared to appreciate the (Malthusian) struggle for existence which everywhere goes on from long-continued observation of the habits of plants and animals', as he put it in his brief autobiography. He also knew, as was common knowledge among plant and animal breeders, that individual differences between members of the same species could be passed on to their offspring. All this he put together in his mind to arrive at two crucial inferences. The first, the principle of natural selection, states that organisms are patchworks of variable traits that collectively sum to determine how well any one organism can survive in a particular environment and how many viable offspring it can produce. An organism that survives and has more offspring than others is fitter in that environment, and, in passing its 'good' traits to its offspring, ensures they too will be more likely to survive and have fit offspring in turn. Variation, selection and transmission are often referred to collectively as microevolution.

Darwin's second inference concerned speciation. The microevolutionary processes, played out over long-enough time – and geological time we now know is very long indeed – will, in variable environments, result in population changes that may, if the selection pressures are strong enough, result in populations diverging from one another sufficiently to become separate species. Hence the different species of finch on the different islands of the Galapagos. This process is known as macroevolution.

The big picture Darwin painted is of specific processes that drive the transformation of complex systems in time, which results in diversity, change and order. Descent with modification means all living forms, including humans, are related, no matter how remotely. Best available evidence indicates our nearest living relatives are chimpanzees, with whom we share over 98 per cent of our genes. Chimpanzees and humans have a common ancestral species that lived around five-and-a-half million years ago, and both species in turn share ancestral species with gorillas and all the other great apes, processing backwards in time. We also share a common ancestry with Darwin's finches. It was probably a stem-reptile species, but we have to go back a long time, some 340 million years, to find it. The common ancestry of our phylum, Chordata, with other animal phyla such as Arthropoda (which includes insects), reaches back close to 600 million years ago, a period known as the Cambrian explosion which saw the burgeoning of multicellular animals.

The wider controversy unleashed by Darwin's theory is still with us today. But while biologists might more narrowly argue about the completeness of Darwin's views, no scientist now seriously doubts that evolution occurs or that selection is an important process in evolution. Nevertheless, Darwin and his followers faced a number of problems. One of these was the evidence put forward to support the theory. The fossil record was there, certainly, as was the distribution of organisms in space of which Darwin was such an acute observer. So too were the specific features, or adaptations (see later), which contribute to fitness in specific environments. However, the great mass of genetic and biochemical evidence on relatedness did not become available until relatively recent times, so everything the evolutionists relied on to support their theory was indirect, circumstantial and open to alternative interpretations. The reason no one could see evolution occurring was, Darwin assumed – though some of his disciples differed from him on this – because it was a very slow and gradual process. People just didn't live long enough to experience it directly. Well, that was the position over a hundred years ago. And that was why the moth *Biston betularia* became famous in biological circles. Until the second half of the 19th century this species of peppered moth was mostly white with a few black markings. Then a dark form appeared and quickly spread to become the dominant form of the species in Victorian Britain's industrialized regions. The lepidopterist J. W. Tutt proposed in the 1890s that here was evolution in action. Because the bark of trees in the Midlands had been darkened by newly arrived industrial pollutants, the dark form was better camouflaged against the birds that preyed on them than was the original white form, which stood out from the dark background of the tree trunks. This was not accepted, alternative explanations being preferred for the moth's change in colour; and, it was objected, birds were not major predators of this species. Astonishingly, more than 50 years passed before Bernard Kettlewell set out to test Tutt's claim. Although his results appeared to support Tutt, the adequacy of this work continues to be questioned.

Fortunately, we no longer have to rely on the peppered moths of England as the only, and questionable, evidence of microevolution in action. There are now many reliable and sound studies which demonstrate the actual occurrence of evolution over just a few years. John Endler of California, for example, brought freshwater guppies from the Caribbean into his laboratory and by clever experimentation in artificial streams was able to map changes in the colours of the fish in response to selection pressures in less than ten generations over a period of just six months. Endler also switched guppies about in the wild and was able to confirm his laboratory findings within a year, findings that confirmed specific predictions. The scientific literature now contains similar reports of long-term morphological changes in the limbs of *Anolis* lizards introduced on to small islands in the Bahamas over a ten-year period. Again, the changes were in the direction expected, in line with differences in density of vegetation in the lizards' new homes. A third example comes from a study of short-lived weeds on a group of Pacific islands off the coast of Canada. Significant changes in the form of the seeds of these plants, which altered their range of dispersal, were observed in less than five generations. Finally, even Darwin's own finches now provide us with evidence of microevolution in action. Two of the main differences among these birds are beak shape and what they eat. Peter and Rosemary Grant of Princeton University have shown changes in beak shape develop over just a few years, and have been conducting experiments on hybridization of some of the species and tracking changes in the form and fitness of their offspring. Hybridization may not be common in animals, but botanists believe hybridization has been crucial in plant evolution. There are other good scientific studies in the same vein. Microevolution has now been seen to be happening. The evidence is direct.

Microevolution, however, is not macroevolution, and it has been the lack of direct evidence of speciation that has really enabled anti-evolutionists to make the claim of 'not proven' against evolutionary theory. But that too has now changed. Early in 2001 the journal *Nature* reported a marvellous study by Darren Irwin and colleagues from the University of California on 'speciation in a ring'. So-called ring species originate from a single ancestral species when small, successive migrations round an obstacle come to form a distribution of populations geographically distributed in a ring – in effect, the populations fan out round the obstacle. Neighbouring populations can and do interbreed. But when the ring is finally closed by the populations meeting at the point in the ring opposite the ancestral population, they have sufficiently diverged from one another that they now form separate non-interbreeding populations. In the early 1940s Ernst Mayr described ring species as the 'perfect demonstration of speciation'. However, until the Irwin paper, no unblemished case of ring species had been observed. Some had been reported but the rings were always too broken for the conclusion to be drawn with certainty that here was macroevolution in action. But now Irwin has reported on a perfect ring species. The organisms in question are birds – warblers – populations of which have fanned out to surround the Himalayas. Using both genetic and behavioural evidence, the latter in the form of birdsong, Irwin and his colleagues have shown that the two species in Siberia, to the north of the Tibetan Plain, are both descendants of the original southern population. The evidence for the occurrence of macroevolution is now as direct as you can get.

Another problem for Darwin was a lack of understanding or explanation of the sources of variation and how traits are transmitted from parents to offspring. On the other side of Europe, and coinciding almost exactly with the period of the publication of *The Origin of Species* and its immediate, fairly stormy, aftermath, a Moravian monk, one Gregor Mendel, was experimenting with the selective breeding of pea plants. Mendel was an excellent experimenter and although his work remained largely unknown for almost 40 years, it contained the beginnings of the new science of genetics. This was to solve almost all of Darwin's difficulties, though Darwin himself died over 20 years before the discovery of Mendel's work and never knew of it. What Mendel showed was that certain entities – factors, as he called them – are transmitted from parents to offspring in regular, lawlike, patterns. Sometimes these are expressed in the structure and function of the offspring, and sometimes they are not. Yet when the latter is the case, the unexpressed factors are not lost but continue to be transmitted to the offspring of the offspring, and in turn to their offspring, and may be, indeed often are, expressed in later generations. Furthermore, these factors do not combine in a blending manner, which was the dominant idea about heredity in Darwin's day. The possibility of blending was a real bugbear to Darwin because if it did occur, variation would be expected to decrease with the passing of the generations. But not only did Mendel show that blending does not occur, he demonstrated the manner in which recombination occurs, which maintains and even increases variation. Because of the mechanisms of cell division and the way in which these factors are randomly partitioned between sex cells in sexually reproducing plants and animals, we now know that sexual reproduction is a biological engine for generating variation on a truly astronomical scale. With the exception of identical twins, it is most likely that no two members of any sexually reproducing species are, or ever have been, genetically identical. Adding to the variation are the complexity and fluctuations in the environment of development during the period in which the genetic instructions in the fertilized egg become expressed as the attributes of the whole organism – the phenotype, in the jargon of biology.

Following the birth of the new science, what Mendel had referred to as factors became known as genes, and genetics blossomed into one of the most successful sciences of all time. The location of genes in the cells, mainly on structures called chromosomes, was established, as was the precise behaviour of chromosomes during cell division. The variation of genes and changes in gene structure – mutations – became known. So did the linkage between genes and how gene expression is affected by the presence of other genes, known as epistatic interaction. Many other aspects of genes were discovered, culminating in the uncovering of the structure of genes as DNA, and how the linear structure of DNA is transformed into proteins, one of the great triumphs of 20th-century science. Recent years have seen the discovery, among so many others, of deeply conserved DNA complexes known as homeoboxes, which persist in creatures as different as fruit flies and humans and play a vital role in development. The history of genetics is the history of an ever deeper understanding and explanation of the similarities and differences between all forms of life.

The modern synthesis, as it became known, which laid the foundations for linking the new science of genetics with the Darwinian theory of evolution by selection, began in the 1920s and 1930s. Different forms of selection acting upon populations of individuals with different genetic and phenotypic structures were described and analysed using multi-dimensional models called fitness landscapes. The conditions under which speciation events occur became better understood, as did the degree of molecular variation in natural populations. One hundred and forty years after the publication of *The Origin of Species* the theory of evolution has matured into a science of increasingly complex detail, with many facets to it, and supported by empirical findings at the level of molecular biology, individual organisms and populations of organisms. The heart of the theory, nonetheless, remains variation, selection and the transmission of selected variants.

While speciation was Darwin's main concern, as the title of his 1859 book makes clear, explaining another characteristic of life – adaptations – was almost as important. Adaptations are those features of plants and animals, often quite complex and unexpected but sometimes beautifully simple, which, once you have figured out what they 'do' and what they are 'for', make you exclaim with delight. What you are looking at is functional design between an adaptive trait and some aspect of the world. So, when Tutt suggested the dark form of the peppered moth was a result of selection for better camouflage in an environment of darker tree trunks, he was saying colouration in that species was an adaptation 'for' concealment.

Adaptations form connections, or linkages, between organisms and the world they inhabit. The connectedness of adaptations is the reason for the ancient view that there is a harmony to life, the forms of which have an intricate linkage both within themselves and in relation to the outside world. Aristotle expressed this in his notion of final cause. He argued that all objects in the world had ends or goals – final causes – that are the reason for their existence. They were what things are 'for', and this was especially so in the case of living things. Christian thinkers subsequently used this seeming harmony between life and its world as proof of the existence of an omniscient and beneficent creator. The theologian William Paley's *Natural Theology – or Evidence of the Existence and Attributes of the Deity Collected from the Appearance of Nature* is wonderfully well titled for the argument it propounds, and is the best-known treatise of its kind. Paley famously compared his response to finding a rock while walking in a field with his reaction to finding a watch. The rock would cause him little problem in accounting for its origins, but a watch had an intricate complexity that spoke of intentional design. This must mean it 'had a maker . . . an artificer . . . who formed it for the purpose which we find it actually to answer'. A maker who had a purpose. Well, what went for watches must also go for the colouration of moths, the beak shape of the Galapagos finches and, everyone's favourite example, the human eye. All are evidence of a Supreme Designer and Maker – God – who created living forms with a view to the world in which they were to live. The design of life is God's design.

One of the reasons for evolutionary theory's savage reception from religious quarters was that it provides an alternative explanation, even when adaptations comprised organs 'of extreme perfection and complication', to use Darwin's own words. Adaptations are the outcome of evolutionary processes acting to increase survival and reproductive competence. It is in the combined effects of variation and selection, coupled to a transmission mechanism, that the artificer is to be found. It was natural selection, not God, that Tutt had in mind when he ventured to suggest that evolution was responsible for the change in colour of those moths. The seeming end-directedness of adaptations is not the result of *a priori* knowledge of a creator, but the *a posteriori* consequence of natural selection acting to sift and filter the genes that are the part-cause of the colouration of moths or the beak shapes of finches, adaptations that are most certainly 'for' concealment or the exploitation of food sources, which are the ends or goals of these particular adaptive traits.

The standard biological view, then, is that adaptations are attributes of phenotypes. (The phenotype is the flesh-and-blood expression of the genotype, which is the total array of genes of an individual from conception.) These attributes are not fortunate and sudden accidents but the crafted end result of selection, long-crafted in the case of complex adaptations such as the eye. Selection can only succeed over time if the information for the construction of attributes is preserved across generations. This means adaptations must be part-caused by genes. They must also be related to conserved aspects of the world, to features of the environment that persist, which is how they come to increase fitness. These are the four necessary characteristics of all and any adaptations: they are of the phenotype, part-caused by genes, end-directed, and they must increase fitness. Any creature is, in Julian Huxley's phrase, a 'bundle of adaptations' that collectively sum to determine how successfully an organism can survive and reproduce.

Because the notion of adaptations is an important part of this book, I need to dwell on several points. The first is that adaptations are one of the central conceptual features of biology that make it different from chemistry and physics. Biochemistry, the chemistry of complex molecules, is one mark of biology, but it is, of course, a branch of chemistry. Take a human being and a bat, grind them up with a pestle and analyse the chemical differences and you will find them only marginally different. That tells us something very important about life, but its significance is the greater when you contrast that minor difference with the differences in the functional organizational of the intact phenotypes of human beings and bats, differences which are startling. Assuming our bat is one of the Microchiroptera, small creatures that are insectivores and live in dark places like caves, we will find that bats have two principal clusters of adaptations that make them utterly different from humans. One is they fly, which humans do not. Bat wings involve massive elongation of the bones of the forelimbs, especially the more distal bones of the fingers. There are also changes in the sternum (breastbone), clavicle (collar-bone), scapula (shoulder-blades) and ribs because of the requirement for stout muscles and muscle attachments. The other is that these bats 'see with their ears', not their eyes. Bats are able to fly in total darkness through a tangle of thin suspended wires without touching them. They catch tiny insects on the wing, and are also able to return to exactly the same roosting place in a large and complicated cave. They do this by using echo-location. Microchiroptera have evolved the ability to generate sounds of very high frequency that bounce off objects – thin wires, small insects and cave walls included – and return to the bat making them, which, remember, is moving swiftly through the air. The bat uses the returning signal to locate objects, such as obstacles and insect prey, and thus navigate successfully through a world without light. That is some feat, some set of adaptations. We humans can neither fly nor echo-locate, or at any rate we cannot do so directly. That we can do so indirectly, through the invention of powered flight and radar, is the result of psychological adaptations which no bat could ever match.

Not only do our bat and human differ only a little in their bio-chemistry, they also share many macro-anatomical features, such as the basic skeletal elements of limbs – also shared with birds and reptiles – and the pattern of their vascular systems amongst other things. But as regards the essential batness of bats and humanness of humans – that is, the functional organization of those features of their phenotypes that are the adaptations for movement, sensory processing and neural computation – they are utterly different. And it is that difference that results in bats and humans having completely different lifestyles. That is why adaptations – functional organization – are the essence, if not the quintessence, of biology. But for adaptations we would all be chemists. And I will be arguing that adaptations of a special kind are the things that provide us with a link between the biological and social sciences. Miracles and theological accounts aside, evolutionary theory is the only explanation we have of adaptations. So while my stance might not be quite as large-scale as that of Szathmary and Maynard Smith, I do need evolutionary theory to achieve some kind of synthesis.

A further point to make is that being 'adaptation oriented', which is what this book is, does not mean one necessarily believes all phenotypic attributes are adaptations, or that all attributes that are adaptations are perfect. A number of processes and mechanisms that result in phenotypic traits are known to have nothing to do with adaptive advantage and selection. One example is interacting growth fields during development that result in structures, such as the human chin, that have no function at all. Another is the irrelevant properties of traits that are selected for other reasons. A frequently cited instance of this is the colour of blood, which itself is not an adaptation but merely the outcome of the light-absorbing property of the molecule haemoglobin caused by the iron atoms that form part of its molecular structure. Indeed, some species have a respiratory pigment called hae-mocyanin in which the metal atoms are copper, as a result of which their blood is blue. Well, red or blue, neither colour has adaptive significance. The different respiratory pigments were selected because of their capacity to transport the respiratory gases oxygen and carbon dioxide.

A related example is the correlations of growth and size known as allometric effects. Generally speaking, the larger an animal the larger its brain. A larger brain is simply a consequence of greater all-round body size, meaning the brain of a big animal such as a rhinoceros or elephant is no larger relative to body size than that of a cat or mouse. What probably *is* an adaptation is a brain size greater than would be expected from body size. The brain of a chimpanzee is about two-and-a-half times larger than would be expected from its body size. The size of the human brain is six to seven times that to be expected from body size. That is an extraordinary discrepancy, the significance of which remains little understood and will be discussed further in the next chapter. The point for the moment is that the consequences of allometric change – absolute dimensions – are not adaptations.

As for perfection, there are, again, a number of good reasons not to expect any adaptation to have the best possible design, even if one could know what that is. Huxley's image of a 'bundle of adaptations', which is what any organism is, means that the existence of multiple adaptations places structural and energetic constraints on all of them. Also, driving design relentlessly down one route often has consequences that are deleterious to fitness. For example, vascularized surfaces are a good design for a temperature-regulating device, and in fact serve just this function in many animals. The larger the surface, the more efficient the heat exchanger. However, a large surface also has costs: it might be too heavy, it might be too conspicuous, and driving blood through an extended network raises the costs of pumping it. So adaptive design is often a trade-off between conflicting needs and possibilities. Such constraints are inherent in any complex system whose multiple functions jostle with one another for limited resources. The overall result is a compromise of less than perfect designs, but designs that nonetheless are functionally significant. 'Tinkering', the French biologists Jacques Monod and François Jacob called it, or 'satisficing' in the language of H. A. Simon, the American polymath. Time and resource constraints lead not to botched jobs but to jobs that often make do with a good deal less than the perfect functional design.

Genetic and developmental constraints are another source of less than perfect design. There is widespread acceptance that selection acting on the phenotypic consequences of changes in genetic structure is the source of all functional design. Genotypes, however, are complex and delicately structured. A random mutation that results in improved design for one trait may well have negative consequences for the balance of the genotype and so adversely affect the expression of other traits. The combined effect of these latter on the fitness of the individual may then outweigh any gains in fitness resulting from the functional improvement to the first trait. In that event, despite its improved design, the first trait will not be selected for because of the negative effects on overall fitness. In short, genetic constraints arise through genetic changes occurring in the context of long-established overall genetic structures. You cannot expect changes in the genes for one trait to have no effect on the genes for other traits. Roughly the same reasoning is assumed to apply to the little understood, but undoubtedly complex, developmental processes by which genetic structure translates into phenotypic structure. Alterations in the development of an improved trait may be offset, and hence constrained, by adverse effects on the development of other traits also contributing to individual fitness.

One other reason for lack of perfection must be mentioned since it may have been important in the evolution of human cognition and the emergence of culture. Darwin himself recognized the possibility of preadaptation, whereby an attribute comes to serve a function which is not the reason why it was originally selected. In other words, evolution has wrought a change in function. The result may be a less than perfect instance of functional design, but evolution the tinkerer makes do. Exaptation, which refers to an adaptation that has recruited to some functional need a trait that either did not have any adaptive function originally or evolved for some other use, is a more recent, if similar, formulation. The wings of bats, which are based on forearm structures the functions of which originally had nothing to do with flight, are one such example. The feathers of birds, thought to have evolved originally for their thermoregulatory effects and subsequently co-opted for another purpose, flight, are another widely cited example.

The reintroduction of preadaptation as a matter of theoretical importance, and its expansion around the wider notion of exaptation by Stephen Jay Gould and Elizabeth Vrba in the 1980s, has raised serious argument and controversy among evolutionists, and some real conceptual difficulties. The original, and core, idea about adaptation is that it is a trait whose positive value for individual fitness is the result of its design being consistently 'improved' by selection for genetic changes that bring about the improvements. The important point in the conceptualization is a straight and exclusive line of causal relationship between genes, the trait, the trait's functional value and selection pressures across time. Each adaptation is functionally compartmentalized, sealed off from other adaptations. But if a trait with one function is co-opted to another function on which different selection forces are acting, or if a trait with no function at all subsequently acquires one, the analysis of causation becomes much more difficult because the causal linkages between genes, traits, the traits' functions and selection become crossed. This is especially the case when, as Gould and Vrba accepted, an exaptation might undergo further modification by selection, an occurrence they termed secondary adaptation. If secondary adaptation has been going on for long enough, then, without access to historical sequences of events and given all the reasons why adaptations are likely always to be less than perfect, it is going to be impossible to know whether a trait that adds to fitness by having a specific positive function and design is an adaptation or an exaptation. For this reason, Gould has suggested the rather stark 'aptation' to cover all cases of design. As will be seen in the next two chapters, the issue of adaptation and exaptation is an especially vexatious one when it comes to a consideration of the evolution of intelligence and cognition. Tempted to adopt *aptation* for this reason, I eventually decided I should stay with conventional usage lest the book appear odd and pretentious. So, on occasion *adaptation* and *exaptation* are used separately. But in the main *adaptation* is used generically, in the sense that its use encompasses doubts as to the extent to which exaptation might have been a part of an adaptation's history.

One other point must be made that will become important in later discussions of culture. The mechanisms Lamarck proposed by which harmony is achieved between organism and environment – that is, of adaptive fit – were very different from those offered by Darwin.

Lamarck's very starting point was an error. He believed that all organisms at any time were in a state of near perfect balance with their environments – an image of natural harmony really is central to Lamarck's vision. Since the world is always changing (Lamarck, too, was struck by that trinity of diversity, change and order), organisms must also change. This, he suggested, occurred through two mechanisms. One is often referred to as the law of use and disuse. Changes in the environment impose new needs upon organisms, which lead to new kinds of activities. These activities, in turn, result in alterations in body structures, because using a body part in a particular way leads to growth and change in that part, whereas disuse results in reduction and even disappearance. Such changes are then passed on to offspring, which was Lamarck's second mechanism, usually referred to as the inheritance of acquired characteristics. We now know that Lamarck's mechanisms are not how biological evolution occurs. Darwin's theory was, in its original version, very different. For Darwin perfection was not a starting point. Rather imperfection, a lack of fit in variants and their elimination by selection, is at the centre of his theory. And contrary to Lamarck, for whom evolutionary change was directed or instructed by already impinging change in the environment, Darwin believed the traits that contributed towards the overall fitness of an organism were generated initially by chance, prior to their being filtered by selection as net contributors to fitness, or not as the case might be, and hence passed on, or not, to offspring. In other words – and this really is *the* big difference between Lamarckian and Darwinian theory – for Lamarck there was direct linkage between adaptive needs and adaptive solutions, whereas for Darwin evolution was a more chance-based (sometimes referred to as 'blind') set of processes in which potentially adaptive solutions preceded adaptive need. This distinction is absolutely fundamental. (References above to *The Origin of Species* are to the original version, for over successive editions, in response to criticisms and difficulties already mentioned, most notably a lack of knowledge of how inheritance actually worked, Darwin gradually incorporated Lamarckian ideas into his thinking. His theory became less Darwinian!)

It might also be noted that Darwinism, for this reason, is often caricatured as a process by which anything can evolve from anything else. This is a half truth because variation is constrained by many things, not least the genotype of any organism, how development occurs for that organism, and the structure of the phenotype that is developed. Instead of 'descent by modification' the theory should more correctly carry the slogan 'descent by constrained modification'.

A final and important word on adaptation: it is not an unproblematical notion. It is very difficult to measure and compare the relative fitness benefits of different adaptations. Hypotheses certainly can, and should be, tested. For example, Nico Tinbergen, one of the founding fathers of the science of animal behaviour, conducted experiments which showed that the cleaning of eggshells from the nest by ground-nesting gulls following the hatching of chicks is an adaptation to the perils of overflying predators, the attention of which is attracted by the highly visible broken shells. So we can assume nest-cleaning by these birds probably adds to their fitness, but by how much, and how much relative to their many other adaptations, such as their ability to recognize their own young, is unclear. It is also often difficult to detect the existence of adaptations. Complexity of an attribute normally flags the presence of an adaptation, because NeoDarwinians assume that only the effects of cumulative selection can explain such complexity. So too does an obvious linkage between an attribute of a plant or animal and some feature of their world. Tinbergen hypothesized that nest-cleaning is an evolved, adaptive behaviour because his field observations pointed to a potential connection between visible cues following the hatching of young and the presence of predators. Not all adaptations, though, are either complex or obvious instances of fit, and in the end the proof of some trait being an adaptation must be empirical evidence. As we shall see, this is often extremely difficult to obtain when one is dealing with possible psychological adaptations.

In the end, the claim for an adaptation remains an hypothesis until the empirical evidence is established that links functional design, no matter how striking and complex, to genes on the one hand and increased fitness on the other. These are very tough criteria to meet. However, explanations cast in terms of adaptations without supportive evidence that the traits concerned are indeed adaptations are increasingly criticized as mere story-telling and a violation of the normal requirements of evidence in science. So, tough it might be to prove a trait is indeed an adaptation, but who said doing science is easy?

Alternative theories to NeoDarwinism

It is an oft-repeated story that many ideas and findings that once seemed to be at odds with NeoDarwinism are now accepted as an integral part of the theory, or as additional sources of diversity, change and order over and above Darwinian selection but not necessarily running counter to it. Only one theory markedly different from Neo-Darwinism has appeared in recent years – punctuated-equilibrium theory – and even that can be partially reconciled with much of the NeoDarwinian thesis.

It was inevitable that the enormous burst of activity in molecular biology that followed the discovery of the structure of genes should have had an impact on a theory reliant upon genetics for much of its source of variation as well as for the transmission of evolved traits. The modelling of the modern synthesis in the 1920s until the 1950s assumed that genes are punctate and discrete, like 'beads on a string'. This image had to be radically revised. Genes, it turns out, are structurally complex, almost messy. They are smeared across chromosomes, with large reaches of DNA not coding for anything as far as is currently known. Genes also form complex families of spatially widespread units. They interact with one another in complex ways, and these epistatic (gene–gene) interactions are not the only kinds of interaction genes enter into. Far from being rather dull, inert, passive stores of information, genes interact in dynamic ways with other cellular molecules, including their own products.

Out of this welter of increasing knowledge, then, came challenges to NeoDarwinism. In the 1960s several theorists independently advocated a 'neutral' theory, which, in essence, suggests that many of the changes in genes are due not only to the processes of selection acting on mutation, but also to random changes that are neutral in consequence and may not even be expressed in the phenotype for many generations. If changes are unexpressed, natural selection cannot alter their frequencies, and so some genetic change can accumulate away from the modulating effects of selection.

Shortly afterwards, in the 1970s, the American molecular biologist A. C. Wilson and his colleagues provided strong evidence that changes in protein structures occur at a rate constant in time and unrelated to numbers of generations per unit time; that is, the rate of evolution appears to be about the same for most species (the exceptions being those with rich behavioural repertoires, which is indeed interesting), irrespective of whether they are long- or short-lived. This would not be expected if selection is the only force for change as it filters the products of each generation.

In the 1980s molecular- or evolutionary-drive theories began to appear. NeoDarwinism pointed to the causal force of selection in altering genes and their frequencies in populations. Neutral theory and the Wilson concept of molecular clocks point to additional, perhaps unselected, sources of variation. But the molecular-drive theorists offer a somewhat different vision of dynamic interactions at a molecular level, depicting them as autogenous drivers of changes in gene forms and frequencies. In a sense the causal shape of evolutionary theory has been shifted to include such autogenously generated genetic change as an additional source of variation and intracellular selection. On the other hand, none of these theories challenges the causal force of natural selection. Indeed, they are themselves selection theories of a kind. And neither they nor any other theory challenges the assertion that adaptations, though hedged about by constraints, are entirely a consequence of selection. So what they have done is enrich evolutionary theory by pointing to additional sources of variation.

A similar complementarity applies to complexity and self-organization theory, which has arisen as a serious theoretical position over the last 15 to 20 years. Complexity theory does not, according to one of its chief proponents, Stuart Kauffman, stand in opposition to natural selection. What it does, via an intriguing series of theoretical concepts, is explore the possibility that when the number of interacting molecules, in the case of the possible origin of life, reaches a certain level of recurrent interaction, spontaneous order emerges via a process called self-organization. These complex, self-organized systems are 'poised between order and chaos' and possess just the right mix of characteristics such that they are neither frozen into such strongly stable states that no change can be wrought on them, nor so unstable that change in any form would cause them to collapse into disordered chaos. In other words, complexity theory attempts to describe the conditions required for 'evolvability'. Evolvability is an important concept because it deals with the fundamental issue of what Daniel Dennett calls 'meta-engineering', which is 'the most general constraints on the processes that can lead to the creation and reproduction of designed things'. What, it asks, are the basic conditions necessary for evolution by selection to occur? Whether selection itself could on its own achieve and sustain evolvability is an intriguing and as yet unanswered question. Kauffman doubts it. The laws of complexity and self-organization, he claims, are a prerequisite for establishing the state of evolvability. Perhaps so. The important point to note, though, is that, like virtually all biologists, Kauffman is a self-confessed Darwinian. He does not believe self-organization replaces selection as an evolutionary force. He does believe that self-organization and selection work hand in hand. One caveat should be entered here. The theory of complexity and its models are exceedingly interesting and beguiling. They may well provide a better understanding of the origin of life, and they certainly have interesting things to say about development, which is one of biology's remaining great mysteries. However, complexity theory is a theory about life for which there is as yet no direct empirical evidence. This presents a marked contrast with NeoDarwinism.

The reader who comes entirely new to evolutionary theory may, at this point, reasonably register a note of protest. Why, it might be asked, have I given the impression that there is *the* theory of evolution when there seem to be several. First there was Darwinism, which was then joined to the new science of genetics, new conceptions such as fitness landscapes were added, and now we have other add-ons such as neutral theory, molecular-drive theory and complexity theory. And there are yet other related ideas being pointed to by important people in the field, like homeostatic equilibrium within gene pools, the need for more attention to be paid to flexibility and control at every level of biological organization, and the insistence that ontogeny (individual development) is underrepresented in evolutionary theory. So why keep on talking as if the essence of Darwinism and NeoDarwinism, variation and selection, is still intact as a single theory when so much must be added to bring it up to date, and yet more is knocking on the theoretical door? At what point has a theory changed so much that it really is no longer the original theory? In a hundred years from now historians of science will be able to tell us, but right now we simply do not know how significant and how fruitful these additions to the theory will prove to be. My guess is that many of them are here to stay, with the whole theory of evolution becoming more and more complicated. Well, that should surprise no one, because what the theory attempts to explain is complicated, and the more knowledge we gain, the more complicated it becomes.

The nearest thing to a really different theory to explain speciation is punctuated-equilibrium theory, first published in 1972 by the American biologists Niles Eldredge and Stephen Jay Gould. It has long been held by evolutionists that the fossil record is a poor, partial and incomplete record of the history of life. By its very nature, only a fraction of life forms would have become trapped into the process of fossilization. That is why we so often do not see a complete succession of evolved forms in the fossil record, one species gradually changing into another, such gradualism being one of the characteristics Darwin, it is claimed, demanded of his theory. Well, Eldredge and Gould took the opposite view. The fossil record, they said, is a reasonably accurate record of evolution, if not in detail then in the general pattern it reveals. What characterizes the record is the persistence of species, largely unchanged, over long periods of time (stasis, so called), followed by relatively rapid change over short periods. Evolution really is 'jerky', as the fossil record shows, not smooth and gradual. Hence the name punctuated equilibrium, which 'focuses upon the stability of structure, the difficulty of its transformation, and the idea of change as a rapid transition between stable states', in Gould's words. There is, it seems, the real possibility of common ground between punctuated-equilibrium theory and complexity theory.

A number of possible causes of punctuation – the periods of relatively rapid speciation – have been offered. One, allopatric speciation in small, isolated populations in which chance drift in gene frequencies may become significant, is not dissimilar to the NeoDarwinian notion of the 'founder effect', where small, isolated groups of organisms are statistically more likely to have a higher proportion of unusual genes and gene combinations relative to that of the total gene pool of the much larger parent population. Cataclysmic external events, such as the great asteroid collision that occurred at the end of the Cretaceous period about 65 million years ago, are another possibility. More speculative are breakdowns in the basic constraints that evolve to maintain a delicate balance in the genotype and its phenotypic expression (homeostatic breakdowns), and macromutations in major gene complexes controlling regulatory and developmental processes (the 'hopeful monster' idea first put forward by Richard Goldschmidt).

Some punctuated-equilibrium theorists are less concerned with the causes of speciation than they are with the radical departure from Darwinian gradualism, the fundamental change in the 'shape' of the history of life on Earth. But the really revolutionary idea of punctuated-equilibrium theory is the decoupling of macroevolution from microevolution. The standard process of NeoDarwinism, variation (including all the newly posited sources of variation) and selection leading to adaptation, is not denied. What is denied is that this 'mere tinkering', by which species evolve adaptations, has any causal connection with speciation. That is certainly a radical departure from NeoDarwinism. Another, somewhat less radical shift, albeit still a major feature of this version of evolutionary theory (we will return to it in chapter 3), is that, unlike NeoDarwinism, punctuated-equilibrium theory brings the structure, the architecture, of evolving systems to centre stage. 'The issue is larger than the independence of macroevolution. It is not just macroevolution versus microevolution, but the question whether evolutionary theory itself must be reformulated as a hierarchical structure with several levels' – again, in Gould's words. This is a genuinely important and significant feature of punctuated-equilibrium theory for anyone using evolutionary theory as a bridge between the social and biological sciences. As pointed out earlier in this chapter, architecture lies at the heart of any attempt to synthesize across these areas of science.

While science is a highly creative process when it comes to ideas and to generating variants, this creativity is counterbalanced by considerable conservatism and caution in the reception of new ideas, the selection filters of the process being very hard to pass through. So when punctuated-equilibrium theory was first advanced it was brushed aside as fanciful nonsense by many evolutionists, some of whom were abusive towards it because they thought the purveyors of this new theory were following an agenda that was only partly concerned with science. A quarter of a century on that ill-judged and intolerant response seems to have been tempered in the 1990s by more judicious views. Now standard texts on evolution devote space to punctuated-equilibrium theory, and even conclude that, basically, sufficient evidence is not yet available to settle the matter of the causes of macroevolution. It is also often pointed out that Darwin himself sometimes wrote like a punctuated-equilibrium theorist. 'Many species once formed never undergo any further change'; 'the periods during which species have undergone modification . . . have probably been short in comparison with the periods during which they retained the same form' – these extracts from *The Origin of Species* have a distinctly punctuated-equilibrium ring to them. As Mark Ridley, the British evolutionist, points out, Darwin's gradualism concerned the evolution of adaptations, particularly complex adaptations such as the eye or the mind of *Homo sapiens*, rather than the pattern of evolution. To repeat the point, punctuated-equilibrium theory does not really challenge NeoDarwinism on the existence of, or the processes leading to, functional design, i.e. adaptations. It is in whether the latter is causal in macroevolution that the new theory is a serious challenger.

How good a theory is evolutionary theory?

In the mid 1970s the philosopher of science Sir Karl Popper wrote an intellectual (*sic*) autobiography in which he declared: 'I have come to the conclusion that Darwinism is not a testable scientific theory, but a metaphysical research programme – a possible framework for testable scientific theories.' This statement added quite a lot of fuel to the fires then burning (raging, some would say) within the biological- and social-science communities. Punctuated-equilibrium theory was being pitched against orthodox Darwinism, schools of systematists were at war with one another as to the relevance of evolutionary theory to the matter of classifying organisms, there was a serious attempt to resuscitate Lamarckism, and the social scientists, and some biologists along with them, were outraged at the encroachment of a fairly extreme extension of NeoDarwinism – sociobiology (see chapter 4) – into its territory. In the end Popper retracted his statement – was forced to, some said, since these were lively times – and declared evolutionary theory respectable. Well, this wasn't really necessary. A 'metaphysical research programme' is a more lofty and important feature of any science than is a theory. It is a higher-order conceptual structure. Be that as it may, let's accept Popper's reinstatement of evolution as a theory, and take a few paragraphs to assess the status and achievement of evolutionary theory as we approach the 200th anniversary of its first appearance in Lamarckian guise.

Earlier in this chapter I said a theory must provide a causal explanation of a phenomenon or phenomena, and science must make predictions about events that can be monitored and measured. By these criteria, evolution is a good theory and evolutionary biology is solid science. Every evolutionary textbook now parades a plethora of data, often in quantitative form, which confirm the basic tenet of evolutionary theory: descent with constrained modification from one or a small number of original living forms. Highly conserved molecular forms, such as DNA, RNA and other features of cellular biochemistry, as well as ancient gene complexes – homeoboxes – bear witness to the sharing of ancient ancestry. Even more impressive are the molecular studies of recent years, which show a reasonably exact gradient in molecular and genetic differences between species that correlates well with the phylogenetic distance established by older methodologies using macrophenotypic measurements. And where older methods were uncertain in their findings, molecular methodology now settles disputes because of its precision. Countless lineages in all the major phyla have now been ordered by a variety of methods and placed within the context of geological time. And while occasional findings of species thought to have been long extinct do occur, none have yet turned our ideas upside down. No tyrannosaurus has popped up in the geological strata beyond that 65 million year boundary between the Mesozoic and Cenozoic eras, and human fossils do not appear earlier than two million years before the present. The fossil record remains consistent with the theory.

A much noted feature of evolution is that the nature of the processes certainly does make prediction less easy than in, say, physics or geology. Nonetheless, predictions can be made, predictive studies have been undertaken, and, as already described, evolution has been literally seen to occur. There have also been many extensions of the theory to account for specific phenomena, often with great success. For example, population geneticists can explain, and predict, the upwelling of variation when previously separated breeding populations intermix. Closer to the theme of this book, the extension of evolutionary theory to account for non-human social behaviour has been spectacularly successful, to the point of replacing whole areas of previous scholarship – what the historian of science Thomas Kuhn would have called a paradigm shift. There are many other examples of predictive, quantified science taking place under the umbrella of evolutionary theory.

What of explication in terms of causal mechanisms? Variation and transmission are rooted in known – and known in increasing detail – genetic mechanisms. Selection has been charted through studying how phenotypes interact with one another and with the abiotic world, from the level of individuals, through groups, to species and ecological communities. Such causal mechanisms themselves are used in mathematical models to predict and explain phenomena such as sex ratios in particular circumstances, and even the evolution of sex itself. As already noted, the theory has grown and is becoming ever more complex, but it is doing so as an inevitable consequence of the discovery of more and more causal mechanisms.

There continue, of course, to be mysteries and unsolved questions. Some of these are the biggest problems of all, such as the origin of life, the definitive causes and pattern of macroevolution, and how ontogeny, the development of individual organisms, fits with evolution. And there is the no small matter of applying the theory to ourselves, especially our minds. So the theory may be incomplete, and is likely to stay so for a long time to come. But it looks increasingly unlikely that it will be shown in the future to be completely wrong. When all is said and done, a dispassionate view must surely be that the study and theory of evolution have matured over two centuries into sound, and sometimes astonishingly successful, natural science. We have increasingly powerful explanations for that diversity, change and order in the world of living organisms that we all see from our windows. It is inevitable that the theory should be increasingly used in the attempt to improve our understanding of ourselves. It is also inevitable that mistakes will be made along the way. Yet it is surely better to try, and to learn by those errors, than to ignore such a majestic biological theory. Do we demean and devalue ourselves by exposing our most exalted features to the light of such theory? Of course not. 'There is,' as Darwin said in the closing paragraph to his *Origin*, 'grandeur in this view of life.'

Suggested Readings

Dennett, D. C. (1995) *Darwin's Dangerous Idea*. London, Penguin Books. (An approachable summary of evolutionary theory, though very NeoDarwinian in orientation.)
Ingold, T. (1986) *Evolution and Social Life*. Cambridge, Cambridge University Press. (An anthropologist who is highly critical of simple-minded application of evolutionary theory to the social sciences.)
Kauffman, S. (1995) *At Home in the Universe*. London, Penguin Books. (A readable account of complexity theory.)
Ridley, M. (1996, 2nd edition) *Evolution*. Oxford, Blackwell. (A complete survey of contemporary facts and theories of evolution.)
Szathmary, E., and Maynard Smith, J. (1995) 'The major evolutionary transitions'. *Nature*, vol. 374, 227–32. (A summary of the authors' book.)

2

The Evolution of Intelligence

Culture is a product of human intelligence. That means that if we are to get any kind of biological 'bite' on culture, we must first see what can be made of human intelligence. Now, human intelligence did not spring fully formed into the world some time in the last few hundred thousand years or so. Its origins lie in non-human intelligence and stretch back in time, over hundreds of millions of years, to the very beginnings of all intelligence. We cannot begin to understand our own intelligence, and hence we cannot begin to understand culture, if we do not place it in the context of all forms of intelligence and the source of those intelligences. There is, as will be seen, a great deal that cannot be known, and much that is uncertain. But some minimal analysis has to be undertaken in order to establish a kind of background against which human intelligence can be placed, thus giving us the basis for thinking about culture from a biological perspective. However, the aim of this, and succeeding chapters, goes beyond exploring an evolutionist's-eye view of culture. My intention is to nail two very important issues. The first is the extent to which human behaviour, including that deriving from culture, can be accounted for in terms of genetic causes. In part, this is the matter of genetic reductionism, which many social scientists see as the great threat to their disciplines. Sorting this out properly is absolutely essential if we are really to see a marriage between the social and biological sciences. The second is the rather older and no less problematic matter of how much of what we do, if not all of it, is a consequence purely of our individual experiences, or whether, as Plato argued over two thousand years ago, each of us brings into this world innate knowledge. This is the rationalist–empiricist division, often represented by adherence or opposition to the notion of the *tabula rasa*. Finding answers to both of these depends on how we understand the evolution of intelligence.

We begin with several questions. What is intelligence? Why did intelligence ever evolve at all? How is it distributed among living things? And what, if anything, is unique to human intelligence? These are deeply contentious, difficult and unresolved issues. Some answers can be given with reasonable certainty, but the reader must be clear about what we are dealing with here. These questions are some of the most central and important that we can ask about the nature of the human mind. They address the issues of just how flexible our intelligence is, and what the limits are to our ability to know and, acting on the basis of that knowledge, to adjust our activities appropriately. Some believe the answers will reveal a species of unique and unrivalled flexibility. We can be whatever we choose to be, and can, with appropriate education, know whatever it is we want to know. Human potential is limitless. Others doubt this and consider the human mind in some important ways to be as limited and constrained as those of other intelligent animals. At a fundamental level, how we answer these questions determines what we think it means to be human. As old as Greek philosophy as they are, what this chapter does is cast them in the context of contemporary science. Given that these are very big issues indeed, what follows is of necessity a truncated version of a long and complicated story told elsewhere. The readings listed at the end of the chapter will point the interested reader to the places where that longer story can be read.

Why intelligence ever evolved at all

Let us begin by defining intelligence as a special kind of adaptation that generates adaptive behaviour by altering brain states. This is a dry and rather stark definition somewhat remote from the everyday meaning of the word *intelligence*, but it will be steadily filled in and transformed into something more familiar as we go along. Well, defined in this way, most living things are not intelligent. Estimates of the number of extant species vary, some putting the figure at less than five million, others at anything between 10 and 30 million. Many of these are single-celled creatures, such as bacteria and protozoa, or plants and fungi, all of which lack a nervous system, the organ that is specialized to detect and respond to certain kinds of events in the world. Almost all multicellular animals have nervous systems, and in some cases, the arthropoda for example, these are well developed and centralized nervous sytems. (The Arthropoda is a group of animals, called a phylum, which includes insects, spiders and crustacea. It is an enormous phylum and accounts for about 80 per cent of all animal species.)

Now, it is an understatement to say that not all species of beetle, for example, which make up about half of all insect species, have been investigated to establish whether they do or do not have intelligence; indeed, fewer than half a dozen out of perhaps as many as a million beetle species have been studied. This gives an idea of the scale of our ignorance. We really do have very little certain, empirical knowledge of what intelligence is out there. However, anyone who has studied the small literature on learning in beetles will not be confident that it is lack of research alone that is responsible for so few reports of learning in invertebrate animals. It is much more likely to be the case that intelligence is very thinly spread outside our own phylum. This is an absolutely key matter to which we will return repeatedly over the rest of this chapter.

A simple form of learning, called habituation, by which an animal ceases to respond to non-significant events, has been reported in some coelenterate species (examples being hydras and medusas), which have a very simple and diffuse nervous system. Habituation is also reliably reported in a number of other invertebrate species, notably molluscs (animals like snails and slugs). There are also reports of a different form of learning, called simple associative learning, in a scattering of other invertebrate forms, including flatworms (platyhelminths), earthworms (annelids), various other kinds of mollusc (octopods and squid) and arthropods. Many of these findings, however, are disputed or weak. In general the status of these studies is poor. Honey-bees, however, which are insects, do show associative learning in studies that are of high quality. Indeed, honey-bees demonstrate certain learning characteristics remarkably similar to those of laboratory rats. Nevertheless, it is only in our own phylum, the Chordata, and specifically among the vertebrate chordates (fish, amphibia, reptiles, birds and mammals), that intelligence appears to be widespread; in fact, intelligence is more diverse in form among chordates than in all other phyla. However, it is worth getting the figures in perspective. Vertebrate species number in the region of 50,000. Assuming total extant species to be, very conservatively, five million, and assuming intelligence is really widespread among all vertebrates (which is questionable, and if it were it would mostly take the form of associative learning), the ballpark figure is something in the region of 1 per cent of all species having this characteristic. Assuming these estimates are far too conservative and intelligence is more widespread in invertebrates than currently thought, it would remain the case that intelligence is present in a fairly small minority of animal species. The real diversity of intelligence, and what many consider the interesting forms of intelligence, such as the ability to learn complex configurations of pattern, including visual and auditory discrimination, complex motor skills such as tool-making, and problem-solving and reasoning, appears to be limited to reptiles, birds and mammals, which constitute less than half of vertebrate species. And when you get down to the species that really interest people, the numbers are very low indeed: in the region of 50 species of corvid bird (such as crows and rooks), about 245 species of carnivore (cats, dogs, wolves and bears), some 78 species of cetacea (whales, dolphins and porpoises), and around 180 species of primate (including prosimians, such as lemurs, as well as monkeys, apes and humans). The really smart species are rare. Well, why is intelligence of any kind a minority characteristic, and why are the exotic forms of intelligence of the kind humans have so uncommon?

In order to begin to answer these questions, we have to get down to basics, in this case the laws of thermodynamics mentioned in the previous chapter. Life is characterized by an improbably high degree of order. There are many ways of illustrating this, one of the most common being to point to an inanimate system. Consider what happens when we place a spoonful of sugar in a cup of tea. Of course, at first it sinks to the bottom of the cup, where it continues to form a mass of granules. But almost at once it begins to disperse throughout the volume of the cup. (Stirring with a spoon simply speeds up this process of diffusion.) Different accounts of this phenomenon all point to the initial state, where the sugar molecules are concentrated in one area of the cup, as a highly improbable distribution caused by the initial forces acting on the sugar mass as it is dropped into the tea, whereas the state of being dispersed through the tea is a more probable configuration because there are so many ways for the sugar to be distributed in this fashion. There are many more ways of being disordered than there are of being ordered. This is one way of describing the famous second law of thermodynamics. Now, if a heap of sugar in one part of a cup of tea is an improbable state that occurs only under special circumstances, how much more improbable is the complex order of living things, *which is maintained through time, often quite long periods of time*? Well, the answer is very much more indeed. Work has to be done to maintain this order, and work requires energy. Plants have evolved a complex process, called photosynthesis, whereby energy from the sun is trapped within their chemical structure and then used to fuel their needs. Some single-celled life forms can chemosynthesize. But animals can do neither, so must gain access to the energy of the sun indirectly by consuming plants and thus utilizing the energy conserved in them, or consuming other living forms, such as fungi, that have their own methods of gaining access to energy, or consuming other animals that eat plants (or even consuming animals that eat other animals). Unlike sunlight, which periodically washes more or less evenly across the whole surface of the planet, these indirect energy sources that animals rely on are localized and scattered. Their distribution is patchy. For example, energy-rich fruits are very specifically located, as are the animals, such as wasps and monkeys, that feed on them. In order to gain access to these indirect energy sources, animals must move. This may take the form of mass migrations of herd animals, the more precisely aimed actions of predators tracking prey, or the movement of just a part of an otherwise largely immobile animal, such as the striking, sticky tongue of a chameleon. In each case, the animals must be able to detect the energy sources, and to act on the world in a variety of ways to access them. In short, they must behave, and behave in a goal-directed manner, the goal being energy supplies. For completeness it might be noted that animals also recruit their behavioural capacities to the cause of other major needs, such as preventing injury and gaining access to other localized and scarce resources, including mates.

Most behaviour in most animals is the result of inherited patterns of central-nervous-system structure and function that have evolved because they are efficient means of exploiting needed resources or effective devices for avoiding injury. I am not going to make a song and dance about the background assumption that efficient, effective behaviours are products of evolution and hence are adaptations (see chapter 1). No serious scientist disputes that the brain and the behaviour it generates are subject over time to shaping by a history of selection pressures in the same way as are other organ systems, or that behaviours which improve fitness become widespread and character-istic of a species in the same way as do other species-typical character-istics. As Darwin observed, in this respect behaviour is no different from corporeal structure. The laying down and following of scent-trails by many kinds of insect, such as ants, is an evolved trait in exactly the same way as is the structure of their mouth-parts or the form of their circulatory system. Nest-building by birds is an evolved trait as surely as the lightweight bones of their bodies are an adaptation for flight. Such adaptive behaviours are usually referred to as innate or instinctive in that they have two main causes. One is the information contained in the animals' genes; the other is a sequence of development, from the fertilized ovum onwards, within an environment that allows the genetic information to be expressed in the relevant organs of sensory systems, brain structures, and pathways that control motor responses.

Three points should be noted here. The first is that individual development is a complex cascade of, as yet, little understood processes and mechanisms. Its importance, however, is undeniable. Second, because of the dependence of genes on an appropriate environment and developmental sequence for the information coded in them to become embodied in neural structures and their behavioural outcomes, genes are only ever part-causes of adaptations of this, or any other, kind. This should not be taken as downplaying the important role of genes as crucial repositories of information that can be transmitted across generations. Indeed, in the overwhelming majority of animals, this is the *only* way information is transmitted between animals. Genes are, however, wholly dependent on individual development for their expression.

The third point is that it has become fashionable to decry the terms *innate* and *instinct*, especially among social scientists and especially when applied to our own species. This is nonsense. Just as it is perfectly proper to describe the response of a moth to light as innate or instinctive, so it is proper to refer to, say, the behaviour of ducking one's head and raising one's arms protectively before one's face in response to a stimulus that looms rapidly in one's visual field, as innate or instinctive. It is behaviour that is part-caused by genes, no more and no less. This is not to say such behaviours are always present at birth, or once established cannot be changed, or might not also be part-caused by intelligence. It simply means that if the genes coding for brain structure and function were scrambled at conception, such behaviours would appear with distorted forms and timing, or perhaps not appear at all.

Now, adaptive behaviours – instincts – wholly caused by a combination of genes and development do not in their enactment result in long-term changes of central-nervous-system states. The brain of a moth is not altered following its seeking-out of a light source, nor is your brain changed by your ducking in response to an object flying towards your head. However, some animals – those rare ones we are calling intelligent animals – have evolved the capacity for altering, in ways not yet remotely understood, the state of parts of their central nervous system. Such changes in central neural states are relatively enduring, and become responsible for generating adaptive behaviours in response to the experiences that caused those altered brain states. This is a somewhat fleshed-out version of our original definition of intelligence as a special kind of adaptation that generates adaptive behaviours by altering brain states. But it takes us further because we can now begin to see the connection between intelligence and the ability to lay down memories, which is what these altered brain states are, through individual acts of information acquisition – that is, by learning. And once the capacity for acquiring and acting on memories has evolved, it then becomes possible to see the connection between the evolution of learning and memory with the further evolution of the capacity to sort, sift and recombine memories through some form of internal manipulation of brain states, to give rise to thought and problem-solving.

Why, though, did intelligence evolve at all? What is the significance of referring to it as an adaptation? And why is it confined to so few species of animal? The first two questions are connected, and easier to answer than the third. Let's go back to the first unequivocal demonstrations of forms of intelligence in non-human animals – associative learning (conditioning) and trial-and-error (instrumental) learning – conducted by Pavlov and Thorndike as the 19th century turned into the 20th.

Pavlov was a physiologist who was concerned to establish a body of evidence to support his theory about how connections – associations – come to be made between different parts of the cerebral cortex, the enlarged outermost layers of the forebrain that are one of the major characteristics of mammals. In experiments with dogs, Pavlov paired various stimuli that at first did not elicit salivation with stimuli that did elicit salivation. The former he called 'conditional' stimuli, for which he used the ticking of a metronome in some experiments and bells or buzzers in others. The latter, the 'unconditional' stimuli, were in some experiments the squirting of small quantities of food into the mouths of the dogs, and in others took the form of much-diluted acidic solutions. After repeated pairing of conditional and unconditional stimuli, the dogs came to salivate in response to the conditional stimulus alone, prior to the delivery of the unconditional stimulus.

Thorndike's methodology was different. He took an educationalist's-eye view of learning and was looking for a way to pursue the study of learning on a scientific basis. With cats as subjects, he worked on the assumption that by learning to pull a string that released the lock on the door of the cage in which it was confined, thus allowing it to leave the cage and eat a piece of fish placed outside, a cat used a process of internalized trial and error.

Neither scientist was much concerned with the kinds of stimuli, responses or rewards used beyond their conforming to generic definitions (for example, a conditional stimulus was not to elicit the unconditional response the first time it was presented), and from these rather mundane beginnings a new laboratory-based science of animal learning and intelligence arose. Over the next 60 years, the basic elements of Pavlov's and Thorndike's experiments were varied systematically in countless ways, serious questions were raised as to whether two fundamentally different kinds of learning had been revealed, and a host of related issues were studied, including the effects on performance of variations in stimuli after learning has occurred, so-called generalization. Pavlov's conditioning experiments, while curiously lax as to precisely which stimuli were used on any one occasion, were nonetheless discussed in terms of the benefits of 'preparatory' responding, just as trial-and-error learning was discussed in terms of the benefits of reward or the avoidance of punishment, which were often tied to basic biological needs and drives. And laws were promulgated in attempts to capture the universal features of learning, both animal and human.

All of this was eminently good science, but it made virtually no connection with evolutionary biology. This was not because the people occupied with the new science of learning were in any way different from those working in other areas of psychology. It was because psychology, by and large and from its inception, never had been oriented towards or influenced by evolutionary theory. It was only in the 1970s and 1980s, some 120 years after the publication of Darwin's first book, that a small number of psychologists began to approach questions in the psychological sciences by relentlessly pursuing evolutionary thoughts. This curious blindspot from which psychologists suffered for so long, and to some extent still do, arose from a combination of history and prejudice. One prejudice was the conceptual positioning of the science of mind, from its inception onwards, such that, like other 19th-century life sciences, notably physiology, which is where its roots really are, it was only ever concerned with proximate causes. Another was the old prejudice many social scientists have against evolutionary biology because of its links with genetics. A third was the stultifying fiat of behaviourism, so damagingly influential in psychology from the 1920s to the 1960s, which decreed that psychologists, unlike other scientists, should not have recourse to causal explanations unless the causes were visible to the naked eye. The result was that the growing science of learning and intelligence had no input from the most powerful theory in biology. This self-imposed isolation should have come to an end as the writings of Konrad Lorenz became more widely known in the 1960s, but because Lorenz was seen as no friend to any kind of psychology, it took another 20 years for evolutionary theory to make its mark on the science of intelligence and learning. There is, as argued in the preface to this book, a strong case to be made for the claim that science is not a social construction in any strong sense of the phrase. However, there is no doubting the extent to which social factors enter the history of our understanding of Nature. The strange case of psychology's relationship with Lorenz was the example I was alluding to in the preface.

Lorenz was the founding father of ethology, the alternative science of behaviour, which is in a direct line of descent from Darwin's work. While it has a long history, predating Darwin, it only became established as a science of animal behaviour in the 1930s, with Lorenz's first reported studies. The award of the Nobel prize in the early 1970s to three ethologists, one of whom was Lorenz, was a mark of the prestige it had gained among the other natural sciences. What distinguished ethology from psychology was that while for the latter evolutionary theory was an irrelevance, for the former it was central to their work. Ethology was based entirely on evolutionary premises. Predictably, it was Lorenz who first took an evolutionist's-eye view of learning, so it was not surprising it was Lorenz who pointed in the 1960s to an absolutely crucial feature of learning that 60 years of intensive study by psychologists had missed. This was that learning is almost always advantageous in outcome for the learner. Given the general intellectual climate within psychology at the time, which so strongly favoured a *tabula rasa* approach to learning and intelligence, really to understand the power of Lorenz's observation was, and to some extent still is, to undergo a conceptual epiphany. Lorenz's insight was also the more remarkable because at the time there was relatively little evidence to support what he was saying. At least partial confirmation came with the increasing realization in the late 1960s and 1970s that some things are more easily learned than others, and full confirmation arrived with a host of empirical studies across a number of species, including our own, which demonstrated that not only is learning constrained, but constrained in ways that make learning mesh with other biological characteristics of the learner.

Some of the most powerful evidence has come from work on birds. Field observation has revealed that hummingbirds forage systematically by exhausting the nectar in one group of flowers before moving on to a fresh group, seldom paying repeat visits to already foraged locations until a substantial period of time has elapsed. This makes intuitive 'good sense', because returning too soon to flowers that have not had time to replenish their nectar supplies would be nutritionally unfruitful and energetically costly. It would not be adaptive behaviour. These birds have learned about spatial position and are acting on this learning, much to their benefit. In the 1980s, Alan Kamil and his colleagues in the United States were able to show experimentally that hummingbirds in the laboratory learn a win–shift strategy (broadly speaking, this means that if rewarded for choosing a location on one occasion, they choose an alternative location on the next occasion) far more easily than they learn a win–stay strategy (that is, to do on the next occasion what was rewarded on the last occasion). While win–stay strategies had frequently been demonstrated in previous laboratory studies of other species, the superiority of learning a win–shift strategy in hummingbirds was exactly what field studies of their behaviour had predicted. This is a lovely case of field studies generating predictions that laboratory experiment confirms.

Then there is the marvellous work, mentioned in the previous chapter, on birdsong. It is the males of songbird species which sing, and they only do so if they are exposed during a fairly restricted period in the first year of their development, at a time when they themselves are not yet able to sing, to the song of adult male birds of their own species. This is an astonishing finding. Remember that during normal development young birds are exposed to all manner of sounds, including the songs of other species. And the experimental evidence is clear in showing they must hear birdsong if they are to sing at all. But it is their own species' song, together with the local dialect variant, that they learn. It is as if they come into the world knowing what it is they have to learn. That is about as strong a constraint on learning as you can get. The seeming paradox of knowing what has to be learned goes to the heart of the magnitude of the conceptual revelation that Lorenz's insight, and these kinds of findings, bring to the science of learning.

There is rock-solid evidence of learning constraints in mammals too. For example, work in the United States by Steven Gaulin and Randall Fitzgerald on two closely related species of vole, one of which is polygynous and the other monogamous, has shown that the males of the polygynous species are significantly better than females of the same species at spatial learning when tested in the laboratory in a series of mazes. No such sex difference exists in the monogamous species. Again this makes intuitive biological sense, because field studies have shown that both sexes of the monogamous species are sedentary, as are the females of the polygynous species. The superiority in spatial learning ability of the polygynous males both over that of its female conspecifics and over both sexes of the monogamous species accords well with the behaviour of the polygynous males, who range widely in space as they compete with other males for access to mates. We do not yet know what is responsible for such learning differences, but their existence is beyond doubt, as is the way they complement the general behaviour and biology of the animals concerned.

Such findings came long after Lorenz's insight. Perspicacious he may have been, but his acumen was born of the need to understand behaviour, whatever its causes, in evolutionary terms. Lorenz postulated the existence of what he called innate teaching mechanisms, which guide the learner in terms of what must be learned – wonderfully apt in the case of songbirds – and hence result in the learned behaviour being adaptive. And this explanation can be valid only if the innate teaching mechanisms, the constraints that direct learning to specific features of the world, are themselves products of evolution – that is, if *they* are adaptations. It cannot be through chance that such constraints consistently cause learners to learn what is good for them.

So, the argument so far is that the constraints on learning and intelligence are adaptations. But what about the basic process itself? What are the origins of intelligence in its most fundamental form? We cannot answer these questions with certainty. Intelligence, probably in the form of habituation and associative learning, is either very ancient, if the current phylogenetic distribution of learning is the result of descent from a single intelligent ancestor, or has evolved independently more than once, which is much more likely. Either explanation points to powerful selection pressures for the evolution of intelligence, but powerful only within restricted lineages of animals. For the next few pages, reasoned argument is all we can fall back on, and the reader should be warned that, in science, argument without evidence is speculation, no matter how well reasoned. With heed to that health warning, let us speculate, but with care.

As outlined in chapter 1, evolutionary theory was born out of the need to explain diversity, order and change. Darwin's theory was the first major attempt within biology to deal with what in the 20th century became the norm in science – the study of dynamic systems, i.e. systems characterized by flux and change. Biological systems are transformed in time by the operation of evolutionary processes, and perhaps through other forces such as self-organization. Evolution is change, and change is what drives evolution. So change is ubiquitous, but it occurs at different rates and takes different forms.

In chapter 1 we considered how anyone peering out periodically through the windows of their house will observe change occurring at different rates. Some, such as fluctuating light levels, shifting cloud patterns and the movement of uncontained debris, occurs rapidly and, seemingly, chaotically. Others, such as the growth of foliage on trees, takes place at a more stately pace. Our neighbours, too, change in appearance as they age, but this is so slow a process our perceptions are often insensitive to it. If we see people regularly, they may seem to change hardly at all. Then there is change that is so slight and slow we need historical records, and sometimes highly sensitive instrumentation, to detect it. If global warming is occurring, which seems likely, it has taken a great deal of effort and argument to establish its occurrence. Long-term and widespread weather changes are known with certainty to have occurred in the past, but the beat of the ice ages has to be measured in hundreds of thousands of years. The slow drift of the continents must be measured on an even longer time scale, but where the various landmasses are now is not where they were hundreds of millions of years ago. The cosmos changes yet more slowly. The chemical elements have not always existed. It is likely that only the deepest laws of physics have remained constant and unchanging since the beginning of the universe.

There is another thing to note about change. Sometimes it is unpatterned and seemingly unpredictable, at least to an observer localized in time and space. Minor fluctuations in temperature across a day or week may well be part of a broader pattern of weather change, but to the localized observer the broader picture is inaccessible. Other forms of change have a regularity that may bring conditions back to where they were at some time in the past. The cyclical changes of day and night and the seasons of the year repeat themselves seemingly without end. Other change appears to take one constant direction. The increase or decrease in temperatures with the retreat or advance of the ice ages, though cyclical in the long term, would have appeared unidirectional in the shorter term. This brings us to an important, if obvious, realization. The form that change takes depends on just how localized in time, how short- or long-lived, the observer is. In order to make the point, it must be accepted that any animal, indeed any organism, can be thought of as a localized observer. If this appears to stretch credibility, for 'localized observer' substitute 'an animal with sensory systems that allow it to detect some of the conditions of the world'. Now, if that animal is an insect that has emerged from its pupa in the summer and is to die in the late autumn after laying its eggs (which is a common life cycle), the steady decline in temperature and light levels as summer gives way to autumn will be experienced as unidirectional changes. The birds and mammals that feed on the insects and have a normal life-span measured in years are observers less localized in time. They will experience the seasonal changes as cyclical, not unidirectional. On the other hand, even long-lived humans, before the advent of science, would not individually have detected the slow pace of the ice ages and so would have registered no directional change at all. Science itself can be thought of, less fancifully, as an observer, and the power of modern science can be measured by the extent to which it has escaped the shackles of temporal localization and is thus able to see the ice ages as cyclical and the expansion of the universe over billions of years as directional. So, while change is ubiquitous, the form it takes depends upon both the rate at which it occurs and the period over which it is observed. This 'relativity of change to the observer' is important because what it says biologically, quite simply, is that how change is experienced depends on life-span. As we will see, this gives us at least a part-answer as to why intelligence is a phylogenetically limited characteristic.

The argument must now be brought back to the notion of an adaptation, and to one specific aspect of adaptations that was mentioned only in passing in chapter 1. This is the 'aboutness' of any and every adaptation. By 'aboutness' I mean goal- or end-directedness, which is that feature of adaptations that makes us all, from Aristotle to today's viewers of natural-history programmes on television, exclaim with delight at the seeming cleverness of Nature. When we do that, it is because we are moved by the central aspect of adaptations, which is the way in which they are not just characteristics of animals or plants that do them good, but *a relationship of connectedness between organisms and the world in which they live*. The shape and colouration of many plants and animals is 'about' their blending in with the background, thereby reducing the likelihood of their being detected. The nectar of flowers is 'about' attracting and rewarding the animals which act as pollinators. There is a relationship of fit between every adaptation and something in the world outside the organism exhibiting it. Indeed, strictly speaking, every adaptation *is* that relationship, being the very thing that connects some aspect of the organization of the phenotype with some specific feature of the world.

Sometimes the relationship is discrete and specific. For example, there are moths that respond to a particular frequency of sound by ceasing to fly, withdrawing their wings and so dropping like a stone. The frequency corresponds fairly exactly with that of the echo-location signals of the insectiverous bats that prey on the moths. The tuning of the insects' auditory receptors and the connectivity within the brain that results in their reflexively ceasing to fly, and hence dropping, is an adaptation for avoiding predation by bats. The sound sensitivity of the prey is a precise relationship of fit with the frequency-detecting signal of the predator. That sound sensitivity is 'about' how bats locate prey.

Sometimes the relationship is less specific. The stalking behaviour of predatory cats is an aspect of cats' central-nervous-system states controlling motor activity in the presence of prey that constitutes a relationship of fit with the sensory systems and fleetness of foot of the prey animals. Stalking is 'about' the vigilance of prey. And sometimes the relationship of fit, the aboutness, is more diffuse and multiform. The ethologist Niko Tinbergen and his students documented over 30 major behavioural differences between ground-nesting gulls and kittiwakes, the latter being the only species of gull to nest on cliff edges. Unlike ground-nesters, which, as we have seen, clean their nests of shells after their chicks have hatched as an adaptation for evading the attention of predators, kittiwakes do no such thing. Kittiwakes are untroubled by overflying predators, and so are messy nesters. Their aggressive and appeasing behaviours are different from those of ground-nesters because of the physical characteristics of narrow cliff ledges and overhangs. Kittiwakes also forage differently. It's a long list. Taken together, kittiwake behaviour when nesting constitutes an 'aboutness' of life on cliff faces.

So, an adaptation is some degree of *matching* relationship between the organization of a phenotype and certain features of the world. How that matching relationship is established is by the microevolutionary processes of variation, selection and the heritability of selected variants (transmission), as outlined in the previous chapter, giving rise to differential survival and reproduction. Remember that neither rival theories of evolution, such as punctuated-equilibrium theory, nor 'parallel, friendly' theories, such as complexity and self-organization theory, challenge the standard NeoDarwinian account of adaptations being the products of these fundamental microevolutionary processes. It is the matching relationships – relationships of fit or the goal-directedness of adaptations – that contribute to individual fitness, which, in turn, results in the conservation of those aspects of organismic organization that match to features of the world.

Well, how does this square with the ubiquity of change? The answer is that it is evolution itself which squares with change – evolution *is* change, and is a *response* to change. Evolution is Nature's answer to change. Think of evolution as a kind of sampler of information using as its instruments those microevolutionary processes. The unit that is doing the sampling is a breeding population of a species, in parallel with other breeding populations of that species, and in parallel with the breeding populations of all other species. The information being sampled is the state of those features of the world that are of importance for the fitness of the individuals that make up the breeding population. And the information takes the form of gene frequencies in the gene pool of the breeding population, the gene pool being the totality of genes possessed by all the individuals in the population. The gene pool, in other words, is the repository of information – the memory – for the breeding population. What evolution 'aims' to achieve is the accumulation of an adequate store of information in the gene pool for the parent population to be able to give rise to generations of individual organisms whose aggregate of adaptations – of matching relationships between phenotypic organization and the array of features in the world the population inhabits – meets the requirements for survival and reproduction. This is an informational or cognitive metaphor for evolution. Evolution is a machine that generates knowledge at the species or breeding-population level in the form of genes such that life can be sustained. The metaphor sets up the next step in the argument.

In the absence of individual intelligence, the way the whole system works can be thought of as follows. Because of the way sexual repro-duction scrambles the information contained in the gene pool, every conception comprises a near random sampling of that information. And that is all the information an organism is ever going to receive. That is all the information it will ever get from which it must construct itself and behave such that it survives and reproduces. Following conception, an organism must develop in an environment that supports and, because development is often flexible, contributes to the expression of the sampled genetic information. Through individual development a phenotype is formed that bears, within a range of viable variation, the structures and functions of all other organisms that have sampled genes at random from the same gene pool.

Once again, be clear about development. It is a complex cascade of processes leading from genotype to phenotype that can give rise to some variability in outcome as a result of slight changes in the con-ditions in which it takes place. Technically, this is referred to as evolution for development within a range of environmental conditions and developmental 'noise'. Sometimes the variability can be significant in outcome. For example, whether locusts live as solitaries or in huge swarms is dependent upon the environment of development. How the visual system is tuned is a function of early visual experience, and sexual behaviour can be affected by social events that occur years before sexual maturity. In all such cases the genotype combines with the processes of development within an environment of development to produce a phenotype the characteristics of which are at once typical of the population and individually honed by development to fit specific environmental circumstances.

Development, then, is a way of adjusting the final product that begins with conception. It results in variable phenotypes that fit with slight fluctuations in the environment. In this way, development allows phenotypes to track the environment and hence to adapt to a degree of change in the environment. But in most sexually reproducing animals, unlike plants, developmental flexibility occupies a limited period of life. At some stage it runs out, and what the phenotype is at the end of development is what it will continue to be for the remainder of its life.

Well, then, if that is so, here is a problem. Conception occurs at one time, let's call it t_c (time of conception), then the translation of genotype into phenotype occurs over a period of time, and it is only at a later time, t_r (time of reproduction), that the animal reaches sexual maturity. At that point it is sexually competent and may return its genes to the gene pool, where they are further scrambled and sampled by succeeding generations. The period of time between conception and reproduction, the interval between t_c and t_r, varies from species to species. In some species it is a short period, in others longer. In a small rodent like a mouse it is a few months, which is about 10–15 per cent of a total reproductive life of about a couple of years. In a small songbird it is at least a year, which may be as much as 20–25 per cent of a reproductive life of several years' duration. In a great ape it may be as long as 10–14 years, which is approaching 40–50 per cent of reproductive life. A similar period would probably have applied in our hominid ancestors. But in a world where change is ubiquitous, *any* period of time between conception and reproduction may mean the information provided at conception is 'out of date' by the time reproductive competence is reached, even though that information can be tweaked and adjusted by flexible development to ensure a high degree of matching relationship between phenotype and environment. When developmental flexibility eventually runs out, what then? How, after that, to deal with a changing world?

Surely, one might ask, why the fuss about a short period of time, given the masses of information coded in the DNA of the gene pool, and especially given that, as will be argued, intelligence first evolved hundreds of millions of years ago, meaning there can be no certainty on our part about longevity, developmental flexibility and the kinds of change to which animals had to adapt? Well, the point is that we do not have to know the details of life in the long-distant past in order to see a problem for the main evolutionary programme in some circumstances if it has not evolved individual intelligence. Let's go back to change, its rates and its forms. Certainly it is the case that many features of the world change so slowly relative to even the longest t_c–t_r interval, in creatures such as ourselves, that evolution registers them as stable and enduring in the short run; and in the long run, change, because it is so slow, can be adjusted to by changes in the gene pool that track these changes in the world. The slow drift in temperatures during the ice ages could have been adjusted to in just this way. Indeed, all change, whatever its form, can be adjusted to by the main evolutionary programme and tracked by changes in gene frequency provided the change is consistent and far greater in temporal extent than generational time, that period between conception and reproduction. So, as temperatures dropped over thousands of years, a period far in excess of the generational time of any species, characteristics such as increased thickness of coat or the capacity to store more layers of insulating subcutaneous fat could have, and probably did, evolve. The coat thickness of a bear is 'about' such long-term temperature changes.

Then there is very rapid change, such as temperature fluctuations during a single day. Because they are a daily occurrence, we can be certain they will happen, even though they are unpredictable in the sense that we cannot know what the precise temperature difference will be tomorrow between two specific times of the day. This is a form of predictable unpredictability, but because the periods of stability are so short, evolution deals with it by accumulating in the gene pool genes for the construction of rapid responses, such as sweating and shivering, to allow for adjustment almost as fast as the change itself. Sweating and shivering are adaptations that are 'about' rapid alterations in temperature.

But there is another form of predictable unpredictability that presents a rather different kind of problem. This is change that occurs within the interval between conception and reproduction, that may indeed occur repeatedly within a lifetime, but which is interspersed with significant periods of stability. Food and water resources are not always absolutely stable in location. They shift about, and they may fluctuate in quality and quantity. Exactly what the course of a river will be cannot be predicted by the main evolutionary programme;

neither can the precise position of a food source. In some social species, some members of the group are friends, others enemies. That much Nature can predict, but not precisely who the friends and enemies will be. So what we have here are changes of potentially life-or-death importance, which can occur rapidly, and which may result in conditions that endure for appreciable periods of time before change occurs again, and hence which cannot be adjusted to with any precision by the main evolutionary programme. There can be no falling back on innate behaviours because instincts are adaptations to enduring features of the world, and what we have here are short-term stabilities that are likely to change at any time within a single lifetime. So what does Nature do?

Before answering this question, let's be a little more precise about the kind of events we are talking about. What we can call 'associative relationships' are an important class of such predictable unpredictability. Broadly speaking, associative relationships come in two types. One of these is spatial location – where things are, what is next to what in the world. The other is temporal order and sequence – what occurs before or after what. Temporal order and sequence in the world of ordinary experience (that is, not in the world of the incredibly small, the subatomic world, or the massively large, the cosmic world of galaxies and black holes) includes causal relationships. Now causality is itself a complex subject, the technical twists of which are not something I want to tangle with here. I will simplify things, though not too damagingly, by asserting that a causal, or cause–effect, relationship is a generative relationship in that without the event we call the cause, the event we call the effect would not occur. So the relationship is not merely a consistent temporal ordering of the cause preceding the effect, but the ordering may signal a generative causal relationship.

Now the world is awash with such associative relationships. Everything has a location; many things may enter into causal relationships. And what is where, and what causes what, may change. Of course, some features of the world are invariably located, though not relative to a mobile animal, the very movement of which changes what is where relative to itself. But many other things do change position, irrespective of animal mobility. And while many cause–effect relationships are invariant, not all are. Again, a large class of cause–effect relationships are determined by an animal's own behaviour. Animals act on the world and thus change the world. Hence an animal's action in the world becomes a cause of certain effects in the world. The great Swiss developmental psychologist Jean Piaget liked to say that 'in the beginning was the response'. What he meant was that *it is in movement and action on the world that intelligence has its origins.* He was right. Mobility and action are a major source of changing associative relationships, which form a major class of predictably unpredictable events. Such predictable unpredictability is pervasive and ancient. Being able to track such changes must have either become imperative in some species, or given them an adaptive edge over those who could not track such change.

But we have already seen how the main evolutionary programme has a major temporal sampling problem. Unable to track such predictable unpredictability unaided, it did something rather clever. It evolved in a small number of animals the capacity for an organ system, the brain, to change in response to short-term stabilities, such that brain states have become the part of organismic organization that maintains a matching relationship with particular features of the world. The brain, of course, was already an organ system that received input from the senses and which, in response, 'automatically' generated signals to limbs, wings and other parts of the body such that adaptive, instinctive behaviours occurred. With the evolution of intelligence, the capacity for some part of the brain to change in response to detected short-term stabilities then became interposed between inputs and outputs to alter behaviour accordingly. The automaticity of some behaviour was thus transformed, because these altered brain states that match short-term stabilities must be able to influence behaviour such that it too matches these features of the world.

In short, evolution, as a gatherer of information which it normally stores as gene frequencies in gene pools, evolved a kind of proxy information-gatherer in the brains of some animals. This is why intelligence is an adaptation. What intelligence is 'about' is short-term stabilities.

This, then, is one possible answer to the question as to why intelligence ever evolved at all, and why it should be considered an adaptation. But we still have to account for the phylogenetic distribution of intelligence. Well, organisms without nervous systems simply do not have the machinery that allows part of an organ system (and it certainly does not have to be a nervous system as we know it in animals, but functionally it would operate as one) to track short-term stabilities by altering its state, perhaps repeatedly, and translating that change of state into adaptive behaviour. But that still leaves us with virtually all multicellular animals, which is a great many. Two particular factors in combination probably go most of the way to account for the apparent sparsity of intelligent life forms on Earth. The first concerns the relative costs and benefits of instincts as opposed to behaviour driven by intelligence. As we have seen, instincts are adaptations honed over long periods of evolutionary time. They are effective because they work instantly. An animal that had to learn to freeze when confronted with mortal danger would probably not survive the learning experience. Many instinctive behaviours have to be fully formed the first time they occur, and the intervention of intelligence of any sort would be distinctly disadvantageous. Instincts also win out over behaviour driven by intelligence because they are energetically less costly. The evolutionist G. C. Williams' principle of the economy of information asserts that living things evolve in such a way as to maintain information in the cheapest form possible. The nervous system, which is an aggregate of nerve cells each of which is constantly using energy to maintain a particular balance of ions across its membranes because that is the basis of neural activity, is energetically a very expensive form of tissue. It is cheaper to code information in DNA than in neural form. It is also the case that the interpositioning between sensory input and motor output of neural networks which track change by changing themselves, and then computing those changes into behavioural consequences, requires many more nerve cells than do behaviours that are 'hardwired' by genes and development alone. Intelligent creatures can often track changes of different kinds simultaneously and in parallel. The position of a food source will be signalled by spatial markers, associations of location and features of the source itself. The greater the number of possible parallel state changes of neural networks – that is, the more events that can be tracked simultaneously – and the more complex the computations those neural networks can perform, the greater the number of nerve cells required. In short, if you can get away without intelligence and make do with just instinctively driven behaviours, you are better off in terms of both certainty of effectiveness of action and the energy costs of maintaining expensive nervous tissue.

The second factor to take into account is that most animals are what ecologists call r-strategists, a designation first proposed some decades ago by R. H. MacArthur and E. O. Wilson. r-strategists are short-lived, their life-span is frequently much less than one year, they develop rapidly, and they reproduce in relatively large numbers over just one reproductive period. Put crudely, the r-strategy is a kind of shotgun life-history strategy which copes with an uncertain and changing world by minimizing the chances of change affecting any one animal by throwing at the world lots of short-lived offspring, only a small number of which need to survive for a short time in order to reproduce. For such animals, instinctive behaviour does the job for some; and for those caught out by change, well, there are others who survive.

Another way of thinking about r-strategists is to say they put a lot of eggs in many baskets. The alternative is to put just a few eggs in one carefully crafted and nurtured basket. Instead of a shotgun what you have is a rifle with sights that must be aimed by a skilled marksman. This was termed the K-strategy by MacArthur and Wilson. K-strategists are relatively long-lived animals that develop slowly, have relatively large bodies and reproduce in relatively low numbers. The ratio of length of life to number of offspring is therefore high when compared to r-strategists. Many vertebrates are K-strategists, and almost all birds and mammals are K-strategists. That high ratio of life expectancy to numbers of offspring is what allows for greater nurturance of offspring but also means there is an increased likelihood of encountering changing short-term stabilities, which may offset the costs of intelligence and be the predisposing factor for the evolution of intelligence-driven behaviour.

There are, however, reservations aplenty about the argument I have developed in the preceding pages. Well-developed associative learning in honey-bees is a problem. Strictly speaking, bees are not K-strategists. They do, though, display some fairly remarkable behaviour. The hive is a place where considerable nurturant behaviour of eggs is displayed. (In general, when animals such as insects show nurturant behaviour it is directed towards either the care of the eggs themselves or the provisioning of nest sites such that when the young hatch there is an available food source in the absence of parents.) Honey-bees are also able to communicate positional information about the location of food and nesting resources, information which they systematically alter as time passes and the sun changes position. It is precisely because bees are known to have these unusual abilities, and because bee cultivation by humans makes them freely available for study, that we know so much about them. True, learning is also known to be present in other highly nurturant insects, such as the potter wasp, which shows good spatial-learning abilities. But this just raises further the sense of unease over how little we actually know about intelligence in other invertebrate species. There are, remember, hundreds of thousands of different kinds of bees, wasps, beetles, bugs and other active, busy creatures about which we know nothing in terms of their capacity or otherwise to track change by having some part of their brains alter in response to events about them.

This unease about lack of knowledge spills over into uncertainty about the specific conditions out of which intelligence actually evolved, conditions to which we cannot gain access as the events concerned took place a long time ago. My speculative account of the advantages of tracking change sounds reasonable when set against our own lives and those of the creatures with which we are most familiar, such as our pets or the birds and squirrels in our gardens. We do enter into shifting social alliances, and food and water resources do shift about for mice, elephants and eagles, not to mention bees and potter wasps. But set against the backdrop of ancient animal forms of hundreds of millions of years ago, about which we know little beyond the fossil record, we are without certainty of any kind. The 'logic' of the story may run true – *does* run true – but how well it connects back to real life in deep geological time is quite another thing.

This leads to another possibility, which is not entirely at odds with the scenario painted in the previous pages. This is the idea that intelligence as I have defined it did not evolve as a direct result of the need to track change in the world; rather, it is a property inherent in all, or perhaps just some, neural network architectures. Given the astonishing variety of animal forms in the long history of multicellular life, 'some' might be quite enough. Neural networks were present in many different forms of multicellular animals by around 500 million years ago. They were probably present in the weird and wonderful animals of the Cambrian period, which stretches back a further 100 million years. Perhaps some limited intelligence just 'falls out' of the physiological characteristics of neurons and neural network structure – a nascent property, an almost universal potential, which, under the right selection pressures, blossoms into an active property. According to this view, intelligence is, and was, a seed planted in almost all animals with nervous systems, but its growth is dependent upon the benefits of its realization as an active principle of brain function governing behaviour outweighing its costs. Those benefits would derive from the existence of short-term stabilities, significant for survival and reproduction, that could only be detected and tracked by the evolution of incipient intelligence into its fully functioning form.

If this alternative view is correct, then intelligence is not, technically, an adaptation. As we saw in chapter 1, Darwin recognized the likely existence of preadaptations – that is, attributes that have come to serve functions for which they were not originally selected. Recently, largely owing to Stephen Gould, the notion of preadaptation has been widened to include traits which originated as adaptations for particular functions but which have subsequently been co-opted to different adaptive functions. He refers to this as exaptation. If intelligence is a property intrinsic to certain nerve-net architectures the origins of which lay in the need for information from the periphery of a multicellular animal to be diffusely transmitted to all regions of the organism, then it is an exaptation the adaptive nature of which was recruited from its originally selected state. However, whether adaptation or exaptation, what intelligence does, at bottom, is always the same. It tracks the changing world of short-term stabilities and establishes behaviour appropriate to those changes.

The limits of reductionism

These days biology is big science. And it is big science primarily because of the growth of molecular biology. Following the discovery of the structure of DNA in the early 1950s, the next half century saw advance after inexorable advance in knowledge of genetics at the molecular level, knowledge now being cashed out into molecular genetic approaches to disease, genetic engineering and the technology of cloning. This has been a period of unprecedented growth of knowledge in biology with a profound impact now imminent for the lives of us all. The impression on non-scientists has been of an almost rampant growth in one of the natural sciences. So, when biologists of any persuasion turn their attention to the social sciences, a deep unease is evident among many social scientists. Partly this originates in the reasonable apprehension that outsiders are going to come onto one's own turf and solve the big questions about human affairs while pushing the natives to one side and eventually denying them gainful employment. The apprehension might be understandable but the feared outcome is unlikely.

The unease also stems in part from the incorrect perception that biologists are all one of a kind, namely ravening genetic reductionists whose crass aim is to explain all the complexities of the human mind, including culture, in terms of genes and the forces of genes. This fear of genetic reductionism is unfounded. For one thing, reduction in science is actually a much more complex and technical matter than most people, both inside and outside science, seem to realize. Reductionism is not simply an act of someone from one discipline initiating study in another discipline. There are, in fact, a number of different types of reductionism, each with different requirements, different effects in terms of scientific practice, and different conceptual goals. Only methodological reductionism, which asserts that study should be focused at the most fundamental level of a complex phenomenon, would hold that while the questions asked should not change, social scientists should stop directing their empirical enquiries at human activity and start retraining as physicists – genes, after all, are just complex chemicals, and methodological reductionism does not rule that genes have special status. Of course, there is neither practical nor conceptual power to this form of reductionism, and even in a world where research funding is limited, no one is going to cut off empirical study of human activity and culture on these grounds. We simply have no idea how to study the biochemical differences that underlie different ethnographies, for example. Even if we did, there would simply be no reason to focus research on the biochemistry rather than the ethnography – unless one had a good theoretical reason for doing so, in which case it would be a different kind of reductionism.

The ideal of a unified science actually runs counter to methodological reductionism. As we saw in chapter 1, a form of reductionism called intertheoretic reduction is considered by philosophers of science to be an important part of a unified science. Now whether theory reduction can ever be completely or only partially achieved, and whether it is even a good thing to aim for, is irrelevant to the practice of science at all levels. If the ultimate aim of a restricted theory reduction is, for example, to explain a theory of human warfare in terms of the neurochemistry of the orbito-frontal cerebral cortex (the part of the brain which neuropsychological studies have associated with violence), we had better have in place a complete science of human warfare in which empirical studies of conflict at the group level mesh with appropriate cultural and economic theory, a complete science of the psychology of violent behaviour, and a complete science of the human brain. This is because theory reduction requires properly constructed theory, and that in turn means theories that are empirically supported *at every level*. Put simply, a unified science requires all science to be done. It fails if the science is restricted to just a few arbitrarily chosen levels. The unity of science rules that no science can or should ever be ruled out of existence.

Then there is explanatory reductionism. This is the intuition – and it is never more than that – held by many natural scientists that somehow explanation is most powerful when it is cast at the most fundamental level of a complex system or phenomenon. I refer to it as an intuition because there really is no reason to assume that an explanation of violent behaviour cast in terms of brain function is superior to an explanation cast in terms of psychological mechanisms, nor that either is preferable to a causal explanation given in cultural terms. Explanatory reductionism is also weakened by reductionists who think like this practising what one can call 'localized' reductionism. Thus psychologists will, if they espouse explanatory reductionism, home in on basic psychological mechanisms as the 'proper' level of explanation of, say, economic behaviour. Physiologists are apt to think that neuroscience is the appropriate level of explanation of basic psychological mechanisms, and some biologists are reputed to think that the power of genetic explanations cannot be surpassed. Apart from expertise, of course, there is no basis for localizing reductionism in this way. It is inconsistent with the basic tenet of explanatory reductionism: if explanation is going to be reduced, it should be reduced as far as possible to chemistry and physics. There should be no halfway stations. Brains and genes are not privileged levels of explanation. This kind of argument, however, is not the most powerful that can be adduced by social scientists against explanatory reductionism. The evolution of intelligence, whether as adaptation or exaptation, provides a much stronger reason to reject even the possibility, never mind the desirability, of reducing explanations of human behaviour to genetics.

Consider again the argument in all that has come before in this chapter. Intelligence, remember, is a proxy information-gaining system. It gains information, information that is important, that is hidden from the main evolutionary programme because it concerns events that change too fast relative to the rate of information gain of the main programme. Remember, also, that intelligence is a 'solution' of last resort. Hardwired behaviour is cheaper, more reliable and quicker to produce within the lifetime of an animal. Put in other terms, if the main evolutionary programme could generate adaptive behaviour without having to have recourse to intelligence, it would. But it cannot always do so, so in some cases it invests responsibility for generating adaptive behaviour in evolved, intelligent brains. And that is the important point. *The evolution of intelligence switches the focus of causation of those behaviours that are driven by intelligence from genes and development to the processes and mechanisms of intelligence.* The 'logic' of the analysis of evolved intelligence tells us that the behaviour of intelligent animals cannot be understood purely in terms of genes – if it could it would be a contradiction; because then intelligence would not have evolved. This applies as much to humans as it does to any other animal.

This really is the whole point of everything that has come before in this chapter. There is no point in writing on culture if one is not going to carry social scientists with one, and the faintest whiff of explanatory reductionism is going to set them, quite properly, against a biological approach to their subject. Culture has long been the legitimate object of study by social scientists using social-science methodologies and cast theoretically within social-science concepts and constructs. Now biologists are beginning to explore ways of understanding culture within the context of the natural sciences, and this too is entirely right. But what has to be understood by the biologists is what the limits are to that understanding. Culture is a product of human intelligence. As such, it is beyond the limits of genetic reductionist understanding. It is not, to repeat, a matter of genetic reductionism being undesirable. That is irrelevant, because genetic reductionism is simply not possible, no matter how much some might wish it were. Explanations in terms of genes and genes alone is a fear no one need have.

Intelligence unlimited?

The message of this chapter so far is that it is in the processes and mechanisms of intelligence where we must look first for an explanation of culture. That does not mean we can forget about genes. We will quickly realize we cannot do that when we take a closer look at human intelligence as it is currently understood. The opening paragraph, remember, pointed to the *tabula rasa* as one of the key issues to be resolved, and that is what we turn to now.

First, a minor digression into terminology. In 1904 a French psychologist, Alfred Binet, was appointed to work on a committee that had been set up to investigate ways in which children who needed special help in French schools could be detected early in the educational process. The consensus reached was that some objective diagnostic measure of ability was needed, and Binet set about devising such a tool. Thus was born the intelligence test. Despite serious reservations by some psychologists, almost a century of psychometric study has led to two conclusions. The first is that there is some quality of the human mind that we can call general intelligence, or *g*, which loosely maps onto the commonsense idea of a dimension of cleverness. Some people are clever, smart, highly intelligent; most are about average; and some are not very bright at all. There is probably some form of generalized ability that permeates all cognitive functioning. In intelligence tests,

THE IMAGINED WORLD MADE REAL

which are made up of a variety of subtests, some verbal, some involving the ability to manipulate spatial relationships, others drawing on arithmetical skills or reasoning ability, *g* manifests itself as a correlation between the scores of any one person on these different subtests. The British psychologist who drew attention to this, Charles Spearman, also noted that the correlation is far from perfect. This, he suggested, is because human intelligence also comprises a set of specific skills, particular and discrete cognitive abilities, which he called *s*. Just what the balance is between *g* and *s* has been a controversial matter within the psychometrics community ever since. In the last couple of decades, the American psychologist Howard Gardner has developed the popular theory that intelligence actually comprises eight quite different forms, and that *g* is a doubtful entity. Needless to say, there are also powerful supporters of *g*. The distinction between general and specific intelligence is not an issue for us here. The rise of cognitive science, especially developmental cognitive science, which studies how cognition develops from birth to adulthood, has tended to emphasize specific cognitive skills, such as the development of motor control and the growth of language in a child. The way the word *intelligence* is used throughout this book is in line with the cognitive-science approach to specific forms of cognition and neutral with regard to *g*. In part this is because if one wants to put human intelligence in the context of the intelligence of other animals, the notion of individual differences along some dimension of cleverness is of no help. We all know there are smart dogs and dumb dogs, and clever chimpanzees and chimpanzees that are less bright. But such individual, within-species differences do not form a dimension that can be extended across species. In order to do that, one must focus on the existence (or absence) of specific cognitive skills. Associative learning is a form of intelligence that is widespread among vertebrates. Language is a uniquely human cognitive trait. Of course, as is evident in all that has gone before in this chapter, generically *intelligence* is the word used to refer to that class of adaptations that tracks changes in the world and generates adaptive behaviours in response to such change. There is nothing to be gained by trying to link this biological view of intelligence with *g*.

Returning, then, to the main line of argument, there is no question but that intelligence is constrained. There are too many empirically cast-iron studies that show this. In addition to studies of birds and voles, laboratory studies of rats in the 1960s unexpectedly led to the discovery that these animals are able to associate tastes and smells with the stimuli that arise from being poisoned, whereas they cannot associate cues of spatial location with being poisoned. Rats are nocturnal omnivores with poor vision, so here again we see how learning fits with the general lifestyle of an animal. Quail, on the other hand, are daytime feeders on dry seeds, so it came as no surprise when it was shown that, unlike rats, they can learn to associate visual cues with illness. This is yet another case of constrained learning. While we shall examine specific instances in a later chapter, in general human intelligence seems to be similarly constrained in that we have a variety of cognitive skills, such as language learning, which differ from one another (and often also from those of other animals) in terms of their developmental profiles, those parts of the brain with which they are most involved, and in terms of their functional and computational demands. In all these ways, learning a language is different from, say, learning to recognize faces.

There is also powerful theoretical argument to support the evidence from experimental studies. It goes roughly like this. If learning and intelligence were entirely unconstrained, three things would follow. First, there would be only one form of learning. Language would be acquired by the same mechanisms as, say, learning to ride a bicycle or causal relationships. This must be so because the presence of different mechanisms to learn about different features of the world means that constraint is present. But it would be biologically wasteful to invest in a specialized, and otiose, learning capacity to run alongside a general-purpose intelligence that could learn anything, including, presumably, that which the specialized learning mechanism was primed to acquire. In any event, there is no evidence to support the existence of general and specialized mechanisms running in parallel (which is not to be confused with the coexistence of g and s).

Second, within the limits of the sensory and motor abilities of an animal whose learning is driven by a general-process device that could learn anything, everything that could be learned would be learned. Consider again simple associative learning – what goes with what in space and time. Even within the impoverished intellectual world of a rat or quail, the variety of objects affecting the sensory system is large. Just think of the innumerable stones, protruding roots, rotting fruits and other kinds of edible and inedible objects with identifiably different sensory properties for rats and quails that would be encountered on an average foraging excursion covering a few hundred metres. If everything encountered were learned in every detail relative to the entire detail of every other object encountered and then stored in some giant associative network, each animal would end up with encyclopaedic knowledge of every knowable feature of its world. It would be the equivalent of me knowing every feature of every house, garden and person in every street I have ever walked, cycled or driven down. Forget quibbles about going too fast on a bike or in a car and think of a very slow stroll along a road. If I were to learn everything that flooded in through my senses, that stroll would take a very long time indeed. I would be learning about the elbow of the person at number 44, and how that related to the state of repair of the lower gate post at number 57, and just what cloud cover there was at the moment I noticed that gate post. And that is the third point that follows from the second. If learning were unconstrained, not only would we learn everything, but it would take a very long time. Learning everything is an extremely slow and ponderous process. And that is not what learning is like. It is quick, and it has to be because that is precisely why intelligence evolved in the first place. A random walk by an intelligent animal, rat or human, through the vast search space of all things that could be known would be unlikely to result purely by chance in the acquisition of the information that the animal needed in order to survive.

Researchers in artificial intelligence have been faced with essentially the same difficulty, which they refer to as the frame problem or the combinatorial explosion. Artificial intelligence must be constrained for the same reason that all natural intelligence must be constrained. If it is not, it becomes cumbersome and stupid. Being constrained means being pointed to specific places, or perhaps to specific entities in specific places. These are Lorenz's innate teaching mechanisms, and, of course, constraint is a negation of the *tabula rasa*. We come into this world with innate knowledge, and it is the innate knowledge which does the pointing. The slate is already written on, and the only things that could be doing the writing are genes driven by the main evolutionary programme.

There are psychologists who continue to adopt what is, in essence, a *tabula rasa* position by maintaining that it is development that does the writing, not genes. There are very strong arguments against this view. The linguist Noam Chomsky's 'poverty of the stimulus' argument is one of them. This maintains that one of the most striking features of human cognition is that the output is richer than the input – that the input is impoverished relative to the output. This means that the input is being added to, and enriched by, internal computational mechanisms. So where do these internal mechanisms come from? To answer this question with 'development and only development' is simply to be evasive by having recourse to a regressive explanation in terms of a sequence of unspecified and seemingly magical events associated with development. But the regression backwards in time is not infinite. It ends with conception and genes. That is where development begins. You can't escape from genes by waving the flag of development. Also, development itself is subject to the poverty of the stimulus argument. A sequence of environmental inputs, be they stimulation from the environment via the senses or inputs such as nutrition, is not enough to explain the enriched product of development. There are other factors feeding into the process of development, and these, of course, are genetic.

Another reason for rejecting the developmentalist *tabula rasa* view is that it leans far too heavily on similar developmental experiences in order to account for the high degree of similarity between people in their cognitive skills. We all acquire language, arithmetical capacity, the ability to attribute physical and social causes, motor skills and much else. These are, unless pathology intervenes, identical cognitive abilities, expressed in the same developmental sequences in all people, irrespective of culture. Yet individual experience is a massively chance-based affair. To assume species-specific cognition can be explained by identity of developmental experience is, politely, unfounded. In blunt language, it is absurd. What the philosopher of mind Jerry Fodor calls 'the new rationalism', which is Platonic rationalism buttressed by modern scientific understanding, ranging from the very existence of genes to invariant developmental events and functional utility, tells us that all intelligent animals, humans included, are possessors of innate knowledge.

At this point, let's step back a bit in the argument to the central claim that intelligence evolved because it can gain information about short-term stabilities that the main evolutionary programme cannot see. Remember, the formulation is one of predictable unpredictability, because only the predictable can be detected and dealt with by the main programme through the evolution of the generic adaptation, intelligence. So, in the example 'Who is a friend and who is a foe?', what is predictable is that this is a problem for a social species, but the precise identity of friend and foe is the unknowable – the unpredictable – element for the main programme. So the latter primes the intelligence of a social species, such as ourselves, by pointing to the identifiable features of individuals in the social group as one of the things to be intelligent about. The short-term stabilities of social alliances are identified by the individuals who form them, and the prominent features of the face are the primary basis for establishing identity. Humans are highly sensitive to faces, with facial attention and recognition having a relatively invariant pattern of development and being lodged within a specific part of the brain. The same cannot be said of elbows or ankles. We are very good at facial recognition, which bears the classic signs of constrained learning. This is an example of one aspect of human intelligence having a particular relationship to the main evolutionary programme. Intelligence acts to support it by finding out what is beyond the reach of the main programme. Put in other words, it finds out what the main programme tells it to find out, because the main programme cannot do the finding-out on its own, and hence has evolved the proxy finding-out of individual intelligence. This is the essence of the relationship between evolution and intelligence. The rat is 'told' by the main programme to ignore spatial location when it comes to understanding the causes of its being poisoned and to concentrate instead on tastes and smells as indicators of the source of the illness. The quail is 'told' something different. And what each is told is told in the context of the totality of the biology of the species – rats are nocturnal, have poor vision and acute olfaction, and so on. So, the essence of the relationship between evolution and intelligence is that intelligence is closely nested within the general biology of a species. That is, it is nested within its genes, its normal environments of development, its life-history strategies, and the host of species-typical adaptations that make up the life-history strategies, including behavioural instincts.

It is this embedding of intelligence within the biology of an intelligent animal that I am emphasizing. This was Lorenz's great insight. Intelligence, the instrument of nurture, hinges upon nature. All nurture has nature, and that must mean it is always deeply involved with genes. It is genes that set the constraints that point intelligence to specific parts of the search space and which say to intelligence, 'That is what you must learn and think about.' So, while genetic reductionism is imposs-ible to achieve in any species that has evolved intelligence, it is equally the case that all intelligence can only be fully understood within the context of whatever genes are priming that intelligence. *Intelligence is caused by genes; the consequences of intelligence, however, cannot be reduced to genetic explanation.* This simultaneous independence from and dependence on genes makes intelligence quite unlike any other feature of living things, with the exception of the immune system (which is none too surprising, since the immune system is the one other organ system that tracks events, albeit events confined to the internal environment of an organism, by matching its own states to those of its world). In both cases, what we have is an interweaving of causes of such complexity that it has given rise to endless dispute and misunder-standing.

While no one who knows the scientific literature on learning would now deny that intelligence is always constrained, there remains fierce disagreement as to the site of those constraints, and hence the possibil-ity once again of saving the *tabula rasa* position. At this point the newcomer to this corner of science might ask why people are so intent on saving the notion of the *tabula rasa*. The answer is habit and custom. It simply is no exaggeration to say that, in Fodor's words, 'most cognitive scientists still work in a tradition of empiricism and associationism, whose main tenets haven't changed much since Locke and Hume' (two of the great British empiricist philosophers of the 17th and 18th centuries – the *tabula rasa* was Locke's brainchild). This version of the save-the-*tabula-rasa* argument goes like this. OK, intelligence is in some sense an evolved trait, even if only very indirectly, and the evidence for constraint is now undeniable. But there are two possible explanations of where constraint resides. The one is that it lies in the actual computational mechanisms of learning, i.e. the neural networks that underlie, say, facial recognition and language acquisition, are different. The second possibility is that the learning mechanisms are the same, that is to say the mechanism is general – this is a learning mechanism that could learn anything (and there are candidate networks for this) – and the constraints lie in non-learning adjuncts, such as motivational and attentional mechanisms. These are what are doing the pointing, but the intelligence itself is truly general, hence, in this sense at least, conforms to the *tabula rasa* notion.

Well, maybe. At present we have no way of distinguishing between these two possibilities because we have not the faintest idea as to the neurological basis of learning and intelligence, or, indeed, of attention and motivation. However, the claim that the second possibility saves the *tabula rasa* is nonsense. The outcome of intelligence, its product, is what matters, not the microstructural detail of neural networks. And the outcome is constraint dependent. It really doesn't matter what the slate is made of or what instrument has done the writing. What matters is that the slate is written upon. In other words, intelligence is more than just the computational mechanisms; it must always include the constraints of attention and motivation. In this sense, intelligence is never unlimited. This does not mean that human intelligence is not uniquely flexible. It does not mean that, often with a lot of work, we are not able to acquire skills and do things for which evolution has not directly equipped us. The written word was invented just a few thousand years ago and, until this century, the skill of reading was confined to narrow sections of literate cultures. There is no way reading could be considered an evolved cognitive skill, yet people can, of course, be taught to read, though it needs careful tuition. Contrast that, though, with the far greater facility, almost insouciance, with which we learn to speak. Pathology apart, everyone speaks their native language competently. Illiteracy, by contrast, exists in varying degrees in different parts of the world. Indeed, in some parts of the world, illiteracy is the norm and sits alongside universal linguistic competence. Unless reared in total isolation or with brain damage, nobody slips through the language net, whereas people often do slip through the net of literacy. The point is that while the flexibility of human intelligence is undeniable, the differences in ease and facility with which we use our intelligence for different tasks is equally undeniable. So our intelligence is flexible, but it is not equal to all tasks. Some things do literally come naturally, and it is doubtful that our intelligence is without limits.

Fodor poses a problem

Fodor is a philosopher of mind, specifically a philosopher of cognitive science, who gladly accepts the new rationalist position insofar as it points to nativism, to the innateness of significant parts of the human mind. He welcomes the extent to which it settles the old rationalist–empiricist argument. However, while not doubting that we come into the world with innate knowledge, Fodor does have doubts about its adaptive, evolutionary origins – or, it might be more accurate to say, he has doubts concerning the extent to which we can bring this into the realms of do-able science. Well, Fodor's is one of the best minds in cognitive science and one must take his scepticism seriously. So I want to spend a few paragraphs considering his views.

Many evolutionary psychologists believe the human mind is a set of evolved adaptations, including cognitive adaptations. This is not a story that Fodor is buying. As I understand it, his doubts have two principal sources. The first lies in our having no knowledge at all of the neurological bases of psychological mechanisms and the behavioural repertoires to which they give rise. We therefore have no way at present of understanding how the evolution of x brain characteristics (whatever these are) in response to y selection pressures gave rise to z psychological mechanisms, where x, y and z are unknown – perhaps even unknowable. y is lost in the past. x and z might one day become known, but at present, to repeat the point, we don't know with any certainty yet, or in any detail, how change of any sort in any part of the brain, be it overall brain size, degree of interconnectedness of cortical neurons or anything else, is cashed out into psychological mechanisms and changes in behaviour.

Well, this is correct, although, as will be seen in later chapters, we are steadily gaining in knowledge at a crude level about what brain regions are involved in mechanisms, including some crucial to the human capacity for culture. But we should put our hands in the air and own up to a great deal more ignorance than Fodor is concerned about. This connects with Fodor's second concern, which is his belief that modern humans are psychologically *very* different from modern chimpanzees, and by implication are also radically different from the ape that, around five-and-a-half million years ago, was the ancestor common to both human and chimpanzee lineages. Yet, Fodor believes, neurologically we are not much different from chimpanzees – and, famously, we share with them more than 98 per cent of our genes. This being so, presumably we are not neurologically too different from that ancestral ape either. Now there may be doubts about two aspects of Fodor's formulation of the problem. One is the neurological similarity between humans and other apes, which will be considered in the final section of this chapter. The other is the radical psychological discontinuity he sees between humans and non-human apes. Actually, I am inclined to agree with him on this latter point, but let's play devil's advocate.

There is not much agreement about the cognitive abilities of extant apes and how different they really are from humans, much less any real knowledge about extinct ancestors. The magnitude of the difficulty can be judged by the uncertainty of many primatologists as to what, if any, are the cognitive differences between apes and monkeys. Most simian primates (almost nothing is known about prosimians, such as lemurs) have the ability to recognize and differentially respond to individual members of their social group. Such individual knowledge is essential for acquiring knowledge of 'third-party relationships', which is, according to primatologists Michael Tomasello and Josep Call, a distinctively primate form of intelligence, possessed by no other mammal. For example, an animal may behave aggressively towards another member of its social group, but not if the mother of the aggressed animal is present and the mother is of a higher rank than the aggressor. Such inhibition of aggression is not the result of the presence of any old dominant adult – it must be the mother. Here, then, is an example of primates, monkeys in this particular example, exhibiting understanding that the target of their aggression and the

84

adult female, the mother, have a specific relationship that would result in violent retribution if aggression were not curbed. This is knowledge of the relationship of others without necessary reference to self. Such knowledge is thought to be quite unlike that which can be acquired by the social species of other mammalian orders, such as lions and wolves, which can acquire knowledge only of second-party relationships involving self and others. Some believe that tool use also involves a form of third-party relationship knowledge, entailing as it does understanding of a physical relationship between the tool (a stone, perhaps) and some object (a nut, say) which does not involve self.

Well, this is interesting material. The point of it is that monkeys can act on third-party relationships just as do apes. Monkeys also use tools, though not perhaps with the skill and facility of chimpanzees. Then again, gibbons, which are apes, are not too skilled with tools either. There are some reported differences in the use of mirrors, though the adequacy of the experiments and the interpretation of their findings have been questioned. Nonetheless, what we have here is really quite unclear in terms of psychological differences between monkeys and apes; reading the primatological literature you get a sense that there are systematic differences, but nobody can quite put their finger on it. And one possibility that must be entertained is that perhaps the differences between monkeys and apes are not very great. Well, if that uncertainty and that possibility exist, perhaps the differences between humans and non-human apes are also not great, contrary to Fodor's belief.

Now while psychological differences between ourselves and that late Miocene ancestor we share with extant apes can only be guessed at, those between ourselves and chimpanzees seem so obvious. Yet there are serious scientists who consider there are as many, if not more, similarities between humans and modern chimpanzees than there are dissimilarities. Well, this is an emotive issue. In general, primatologists are not as one in their position on this, but those who advocate only small differences tend to come from this discipline. Psychologists, on the other hand, are mostly of the view that the differences are substantial. The main 'battle ground' (which is just about what it has been) for the last 30 years has been studies of whether apes can acquire language. More recently, what is known as 'theory of mind', which is the ability to attribute intentional mental states, such as knowing and wanting, to others (i.e. to understand that others have minds), has moved centre stage. A small number of scientists believe that clever chimps can acquire language (and it seems they do have to be clever – average chimps can't); most do not think that even the cleverest of chimps can do so. Decisive experiments on theory of mind have yet to be carried out. The picture is the more confusing in that there is little doubt that in captivity, especially if that has involved extensive interaction with humans, chimpanzees do seem to show cognitive skills that are not in evidence in their wild conspecifics. So, a form of enculturation does seem to be within the cognitive scope of chimpanzees. They do seem to have a degree of flexibility to their intelligence, and possibly, like humans learning to read, can acquire knowledge and skills that 'don't come naturally'. Incidentally, the paucity of our knowledge in this area is largely a result of apes in general, and chimpanzees specifically, being a precious and diminishing resource. Solid studies of ape cognition have simply been too few to give us data comparable in quantity or quality with the wealth of information we have accumulated about ourselves. Studying one's own species is so much easier and can be done under a range of circumstances, including pathology. Pathology is not a source of knowledge about ape cognition at all. So, when push comes to shove, while I am one with Fodor in believing that chimps and humans are very different psychologically, I could not put my hand on my heart and say we *know with certainty* that this is so. Perhaps the discontinuity is not as great as Fodor thinks.

It is also sometimes said that five or five-and-a-half million years, the time widely thought to separate us from that common ancestor, is not long enough for the human mind to have been shaped by evolutionary forces into the complex of different mental organs some evolutionary psychologists believe it to be, and which is the main difference between ourselves and that ancestral ape. Well, there are two responses to that. The first goes back to how little we know. We lack the data to make any kind of judgement as to how structurally complex non-human ape minds are. But even if we take the Fodorian position of assuming the difference between humans and non-human apes is great, the second response is to point to just how long five-and-a-half million years is. The arithmetic is simple. Assuming human genera-tions over most of this period to be separated by 20 years – and that figure is probably too high – one million years covers 50,000 generations. It is more than two million years since the first species of *Homo* trod this Earth, which is over 100,000 generations. Five million years allows for 250,000 generations. That is a lot of generations. If the time separating generations is put at a more realistic 15 years and we add in that extra half million years, we are looking at closer to 350,000 generations between ourselves and the ancestral Miocene ape. In the previous chapter studies were described of evolutionary change within a few generations and just a decade or two. OK, the human mind is not the same as the colouration of a fish or the structural features of plant seeds. Nonetheless, a third of a million generations is a lot of evolutionary time. Fodor chooses not to accept the adaptive evolutionary story about the origins of the modern human mind, but five-and-a-half million years can embrace a great deal of evolutionary transformation. So, part of Fodor's rejection rests on our not having the facts about certain things and he points out that progress in subjects such as physiology were made without having to have any recourse at all to evolutionary theory. Likewise, we do not have to have an evolutionary story in order to say we have some understanding of the human mind. Well, of course. That is no great revelation. Indeed, as already indicated, most scientist psychologists have always accepted this to be the case. But overwhelmingly the facts seem to point to the correctness of a strong nativist position with regard to human cognition. That must mean – it can only mean – that genetics is a part of the story. And there is an ineradicable link between genetics and evolution.

The other part of Fodor's doubts lies, perhaps, in the specific kind of evolutionary story that many evolutionary psychologists are trying to tell. This is the story cast in terms of NeoDarwinism and adaptations in Darwin's original sense – traits that have been gradually changed over long periods of time and honed by a history of consistent selection pressures to play a role in adding to the fitness of an organism. But there is another kind of story, a twist to the adaptationist tale, and it is to this that we finally turn.

Human intelligence as adaptation or exaptation

In this chapter we have considered the general argument for the evolution of intelligence, and some of the implications of this view. The specific road from the beginnings of intelligence in simple neural networks to that of *Homo sapiens* is a very long one, is not much known or understood, and is not really our concern. All I have wanted to achieve is the grounding of intelligence within a general evolutionary framework. However, there is a feature of human evolution, better known even than bipedalism, that has come to be associated with human intelligence that must be considered. This is the size of the human brain and its rapid expansion during the evolution of the hominid lineage. The australopithecine apes ancestral to the genus *Homo* had brains roughly equal in size to those of modern chimpanzees, a fact that I will return to shortly because it may have something interesting to tell us. Then, over a period of some two to two-and-a-half million years, hominid brains more or less trebled in size, from around 400 cm^3 to about 1250 cm^3 in modern humans. No comparable change in brain size in just two million years is known. There are many theories, and few facts, as to why the human brain grew in size so rapidly. No aspect of human origins has attracted more speculation, often buttressed by mathematics. Climate change, habitat change, diet change, the complexities of group living and even the notion that humans lived for a time in water (the aquatic hypothesis) all bustle with one another as explanations for why we are the way we are. The fact is that the past is another country and very, very difficult to get to. Fodor is right about this. But for our purposes we don't even have to try to reach it. This is because the most convincing arguments as to how to think about the human brain come from comparing what we know with certainty about contemporary brains. The simple message is that the human brain is not just a big ape brain. It is also in some important ways a different brain. A detailed, but accessible, account of this is to be found in a recent book by the American neuroscientist Terrence Deacon. Only the briefest of sketches can be given here, so for those wanting deeper knowledge, Deacon's work is strongly recommended.

Traditional analyses make a number of suspect assumptions. One is that brain size and amount of intelligence are one and the same thing. This is not true. Brain size scales to body size, though precisely how is not yet settled, without any necessary consequence for intelligence. Compared to mice, elephants have enormous brains. The difficulties of comparing intelligence between such very different animals notwithstanding, there is no reason whatever to believe that elephants are hugely, or even just a little, more clever than small rodents. We certainly know that they are not as smart as we are, yet their brains are around four to five times bigger than ours. Elephants, of course, are just very big animals in every way.

A second suspect notion is that intelligence bears a simple linear relationship to brain size; that is to say, increases in size, reflecting numbers of computing units, nerve cells, computational power, memory or what have you, lead to equivalent increases in intelligence, no matter where you are on the scale. We have, however, no knowledge at all as to how an increase in brain size from 400 cm^3 to 700 cm^3 (the latter being the approximate size of the brain of *Homo habilis*, the first in the line of hominid species leading from the australopithecine apes to ourselves) affected cognitive function, much less as to whether the supposed improvement was comparable to that which occurred when brain size increased from around 900 cm^3 (the estimated brain size of *Homo erectus*, immediate ancestor to archaic *Homo sapiens*) by the same 300 cm^3 to 1200 cm^3 (the brain size of the latter). There is no reason to suppose that equal increases in brain size mean equal increases in any aspect of brain function – disproportional relations between size and psychological function are at least as likely.

A third, and perhaps the most important, of the suspect assumptions is that brain size is a single trait and that size alone is all. There is strong evidence that this is quite incorrect. With changes in size come changes in relationships and in the balance of relationships between different brain regions, changes that must have functional consequences. Change the size of the brain and it is likely that multiple mental traits are altered, not just one. 'Bigger is different as far as cognition goes' is how Deacon puts it. The reason for this is that the genetic and embryological causes of increased brain size have not been a simple change to the genes dictating a simple developmental event, such as 'let the number of divisions of brain cells increase from x to $x + 1$ or $x + 2$', thus doubling or quadrupling brain size across the board. This wouldn't work if for no other reason than that the brain wouldn't fit into the skull. The changes that occurred were deep and widespread to the whole of the head region, and concerned those fundamental genetic units the homeotic genes, the most conserved architects of segmental development, which seem to be shared by almost all multicellular animals. It is clear that not only were the changes widespread in the head region, but that certain parts of the brain have borne the brunt of the changes. It is the dorsal (top) and anterior (front) parts of the brain where growth was greatest, with a change in the growth curve of the brain and of the rest of our bodies such that humans deviate significantly from the typical primate developmental pattern. While we start off, prenatally, pretty much like monkeys and non-human apes, our brains continue to grow for much longer than those of other members of the order Primates. The net result is that brain growth in humans is quite out of line with body growth, and we end up with brains of a size to be expected in an ape with a body some six or seven times larger than that of the average adult person. In other words, our bodies should weigh about 450 kg (that's over 70 st. for the old fashioned) if they were in line with our brain size. But the really important point is the disproportionality in these developmental events. That great grey sheet of nerve and glial (supportive) cells, the cerebral cortex, ends up about twice as large relative to other forebrain structures than it would if human brain growth had been proportional to that of other apes, and three times as big relative to the hindbrain, spinal cord and all other bodily structures. It isn't just that homeotic genes have been involved, but different genes have been differentially altered.

Furthermore, within the cerebral cortex, these deep-rooted developmental changes have resulted in skewed changes in developmental competition for connections with other parts of the brain – those that have shall have more, is the general embryological principle of neuronal connectivity – thus magnifying the alterations initiated by changes in homeotic genes. The end point of all these evolutionary and developmental shifts is that the balance between brain regions directly tied to the periphery of our bodies, the sensory and motor areas, and those brain areas, referred to generally as association cortex, not so directly linked, is quite different from that in apes such as chimpanzees. For example, relative to our actual brain size, visual cortex in humans is almost half of what would be expected, whereas those parts of the cerebral cortex in the most forward regions of our brain, the frontal and prefrontal cortical areas, are about twice as large than would be expected if the brain had simply increased in size equally in all regions.

These developmental displacement effects mean that Fodor might be wrong when it comes to comparing the brains of chimpanzees and humans. Our brains might be as different as are our minds. There may be no Fodorian discontinuity at all. Yes, of course, the basic neurophysiological processes are the same in all animals with nerve cells. Nerves in squid 'fire' in exactly the same way as they do in squirrels. Conduction of nerve impulses across the junctions of nerve cells are the same in rays and rhinoceroses. And it is very likely that the physiological bases for learning and memory are the same in all species that can learn. But the macrostructures of brains, the relative weighting of computational resources and the consequent nature and power of different cognitive functions may vary significantly, and not unexpectedly, when the brains of chimpanzees and humans are compared.

Incidentally – and this was the point alluded to earlier as something to come back to – australopithecine apes had brains of pretty much the same size as those of modern chimpanzees. The dramatic changes seem to have come with the emergence of our own genus, *Homo*, so may well have been compressed into the last two-and-something million years, hence we may be looking at evolutionary change 'compressed' into a mere 100,000 generations, not the third of a million that extend back to the Miocene. However, 100,000 is still a lot of generations.

Now here is the real point of all this. The relative sizes of the brain's functional divisions are the complex result both of widespread competition for space driven by both peripheral constraints (i.e. sensory and motor links) and of shifts in proportions of cells determined early in embryogenesis. 'Though this does not rule out functional and adaptational consequences of individually enlarging or reducing brain structures, it forces us to understand such size variations in terms of system wide effects: *they are not isolated adaptations.*' The words are Deacon's, the italics mine. So if, as is the case, human prefrontal cortex has ended up being about six times larger than chimpanzee prefrontal cortex, this is not solely caused, if at all, by some aspect of prefrontal function – its 'mind' aspect – having been strongly selected for in isolation from all other effects such expansion might have had. The systemic development of the entire brain has to be considered. This does *not* mean that fitness consequences might not have resulted from these systemic changes – they certainly have – but such consequences may not be adaptations in the original Darwinian sense. If the increase in size of prefrontal cortex originated as a consequence of otherwise-caused complex genetic and developmental events affecting the whole brain, which also had the effect of freeing up advantageous computational processing that could be directed towards some specific mental function, such as, say, a theory of mind – if that is so, theory of mind is an exaptation, not an adaptation, which is what Gould argued in a widely cited paper of 1991. Well, perhaps. But does it matter?

The philosopher Daniel Dennett is an orthodox NeoDarwinian who is unphased and unimpressed by the notion of exaptation, because it is likely, he argues, that 'since no function is eternal', every adaptation has evolved out of predecessor structures the functions of which were different, or which perhaps had no function at all. Anyway, exaptations are, after all, adaptations, often themselves fine-tuned by natural selection. Well, Dennett is correct in one sense but not in another. If a psychological trait – for example, competing vigorously for resource in times of scarcity – does have fitness value, Dennett is right whether it is an adaptive trait selected specifically for the usefulness of competitive vigour in a world where resource availability is reduced, or an exaptation in which mating competitiveness has been co-opted to the more general function. Gould's distinction is interesting but trivial. On the other hand, modern humans are capable of some fairly extraordinary cognitive and behavioural feats. Reading, as already pointed out, and writing, and doing mathematics and escaping to the moon when we have finally ruined Earth, are certainly adaptive under appropriate circumstances. Because of the timing of their invention they simply have to be considered as exaptations, co-opted from already evolved visual, manual, linguistic and elementary numerical skills. There is no point in looking for the kind of confirmation of adaptive status that we would with a trait such as ducking away from a looming visual stimulus. Amongst other ways, this is normally done by hypothesizing function through reverse engineering (going backwards from the attribute itself to what it is and was for, and how it came to be the way it is now) on the basic assumption that if it is an adaptation, it is the product of consistent selection pressures over a long period of time. After all, being struck in the face has always been a potential danger, so the evolutionary construction of the ducking response poses no conceptual problems. But reading is different because the function and value of reading have been so recently co-opted that there has been no selection at all – certainly not in Darwin's sense.

Well, that the likes of reading and writing are exaptations is so obvious as not to need dwelling upon. Nobody is going to waste their time telling fallacious adaptationist stories about them. But what if some cognitive skills, such as theory of mind, are actually – and this must be a possibility – only recently acquired cognitive exaptations? In that case we are in danger of overstretching the NeoDarwinist stance by embracing too many mental traits within an undiscriminating adaptionist explanation of the human mind. In the end, our ability really to tell apart adaptations from exaptations is going to depend on future developments in behavioural and neural genetics (establishing genetic links with brain structures, mind structures and behaviour) and on an understanding far beyond anything we have at present about the relationships between neurological and psychological events. And this is important because, in the end, we do want correct causal understanding. That, remember, is what science is all about.

However it turns out, though, it is extremely unlikely that complex psychological processes and mechanisms are not any kind of adaptation at all. As Gould says, 'It would be the most extraordinary happening in all intellectual history if the cardinal theory for understanding the biological origin and construction of our brains and bodies had no insights to offer to disciplines that study the social organizations arising from such evolved mental power.' Quite so. Culture is one aspect of the social organizations arising from evolved mental power, so it is to culture that we now turn.

Suggested Readings

Deacon, T. (1997) *The Symbolic Species: The Co-evolution of Language and the Human Brain*. London, Allen Lane. (The middle section is an excellent account of brain evolution and brain development. The rest is worth reading too.)

Fodor, J. (1998) *In Critical Condition: Polemical Essays in Cognitive Science and the Philosophy of Mind*. Cambridge, Mass., MIT Press. (A recent anthology of critical essays and reviews with a section on evolution and cognition.)

Plotkin, H. (1994) *Darwin Machines and the Nature of Knowledge*. London, Allen Lane. (A more detailed account of the evolution of intelligence.)

Sternberg, R. J., and Kaufman, J. C. (2001) *The Evolution of Intelligence*. Hillsdale, NJ, Erlbaum. (A recent collection of differing views on the evolution of intelligence.)

Tomasello, M., and Call, J. (1997) *Primate Cognition*. (A magisterial review.)

3

The Emergence of Culture

Any declaration that culture is not some single thing, and that no approach or method of study has a monopoly on the understanding of culture, could be read as both a confession of muddle-headedness and an adherence to a destructive pluralism that threatens conceptual chaos. It is a charge that has often been brought against cultural anthropologists and other social scientists studying culture. Yet it is also the position adopted in this book. So what this chapter explains is why a pluralist approach to culture is necessary. It considers the different kinds of thing culture is and briefly surveys the different kinds of approaches made and theories offered in the attempts to understand this strange and most human phenomenon. We will begin where the previous chapter left off, with evolved mental powers and how to think of culture in the light of the main message of that chapter.

That message, remember, is that human intelligence, like all forms of intelligence, is an adaptation which evolved because of the limitations of the main evolutionary programme in gaining essential knowledge about the world; that, for this very reason, genetic reductionist explanations of intelligence-driven behaviour is impossible; that, like all adaptations, intelligence is a constrained and less-than-perfect instrument; and that, like all adaptations, it is part-caused by genes. Remember also that *intelligence* is a generic term. It refers to the bundle of cognitive skills that allows us to learn and think about objects, actions and events, from manipulating objects, through tracking our position in space and on to solving abstract problems.

Whether these skills are adaptations, exaptations or exaptations of exaptations, human cognition, human intelligence in all its forms, has given rise to human culture. In a very important sense, intelligence and culture cannot be separated. But there is one big difference between them. The components of human cognition are mostly assumed to be evolved adaptations. There is a caveat to be sounded in a few pages about this statement. That warning apart, the argument for the present is that whether these cognitive abilities be face recognition, numerical competence, the capacity for imitation, the ability to acquire complex skilled behaviours such as miming actions or making a tool, learning about physical causation, attributing intentional mental states to others, acquiring language, or any of the other cognitive skills – the separate intelligences in the language of chapter 2 – all are evolved adaptations in the wide sense. By 'wide sense' is meant that they have been selected for, either in the strict NeoDarwinian sense or through being co-opted as exaptations, by consistent selection pressures during human evolution. It is the consistency of selection pressures that is important. Present knowledge rules out any understanding of precise historical sequence, so it matters not whether such macrostructural cognitive modules might have been constructed out of similar, if not identical, neuronal microstructural precursors or precursor psychological mechanisms – that is, whether they be labelled adaptations in the strict NeoDarwinian sense or exaptations. What is important is that each form of cognition is an aspect of organismic organization, a trait that matches some feature of the world.

The same cannot be said of culture. Culture certainly does give rise to adaptive responses – many anthropologists have based their careers on this assumption. But culture and the capacity for entering into it are unlikely to be direct products of evolution, in the sense of being unitary traits for which there has been specific selection. Yet even this statement must be modified. At several points in this book we will have cause to invoke 'first culture' – the unknowable point in human history when culture first appeared in our species (and there has to have been such a point, even if we can know little about it) – and contemporary culture. The claim that culture is not a direct product of evolution refers to first culture. The changes from first culture to contemporary culture very likely were driven by direct evolutionary forces, a position that will be explored more fully in later chapters. (Cultural evolution, as distinct from the evolution of culture, has a different meaning. It refers to the assumption that cultures change because of evolutionary forces. This will occupy us in much of the next chapter.)

The difference between the rather diffuse notion of the emergence of culture, and the specific theory that culture arose through evolution by selection is important and needs to be expanded upon. The palaeontological evidence points overwhelmingly to Africa as the continent where early hominids and their immediate precursors, the australopithecine apes, evolved. The first migration out of Africa is thought to have occurred around 1.8 million years ago, although early forms of *Homo* probably remained in tropical climates for a long time after that. It is not until about half a million years ago that there is evidence of species of *Homo* living in temperate and subarctic zones under conditions wholly different climatically from those to which early hominids were biologically adapted. Tool evidence apart, this is the first sign of culture in our genus. This is because survival under these conditions had to have been partly due to the cultural transmission of knowledge and habits to do with protection against the cold. As the American anthropologist Ward Goodenough argues, seasonal changes in weather had to be planned for and acted upon in advance of the arrival of the cold weather. The construction of shelters and the preparation of 'clothing' had to anticipate changes in weather. 'To begin preparing the skins (or other means) needed to protect the body from the cold only after the cold weather arrives is to act too late,' as Goodenough puts it. Precisely what psychological mechanisms were employed in the coordination of such activity is open to question – it is not clear that language had yet evolved 500,000 years ago – but Goodenough's argument does point to culture as the vehicle driving survival under cold conditions. This is evidence of culture, albeit very indirect. And such culture had to be a consequence of psychological mechanisms inherited from the tropical- and subtropical-dwelling hominid ancestors of these ancient humans. To be sure, we have rock-solid evidence (literally) that culture of a kind existed not only in *Homo erectus* but also in the earlier *Homo habilis*, in the form of stone tools, the oldest of which were made more than two million years ago. There is even some speculation that the australopithecines, from which *Homo habilis* evolved, were tool-makers – and why not? Chimpanzees make and use tools and the evidence is that they learn to do so by observing others. So the psychological mechanisms for supporting some kind of culture are very old indeed. But whatever species we are talking about, the mechanisms must always have predated the existence of culture itself. It was psychological mechanisms that evolved first, in response to a host of selection pressures. It was only when some minimal number of psychological traits were in place that culture could emerge.

If one tries to run the counter-argument, which holds that culture was so advantageous to individuals as to be the engine that drove the evolution of its constituent psychological mechanisms, one runs into the problem that the causal arrow is surely pointing the wrong way. Culture is dependent upon a variety of mechanisms. (What these are is the subject of a later chapter.) Without them there can be no culture. *In the beginning* it had to be the mechanisms that evolved first. To argue otherwise is to put the cart of culture in front of the horses of psychological mechanisms. 'Nature' is not prescient, and so could not foresee the favourable consequences of culture. It also could not foresee which mechanisms, in what combination, would support and give rise to culture. However, once some minimal first culture had emerged, with consequences for the fitness of the individuals entering into first culture, some complex interweaving of the forces involved, some kind of synergy between the constituent mechanisms and their outcome, culture, might have occurred, with the advantages of culture then driving further evolution of the constituent mechanisms. Perhaps so. Indeed, probably so. But if we are really to understand culture, and to analyse it in any detail in the light cast by evolutionary theory, we must start with the assumption that the mechanisms had to evolve first. It is from these mechanisms that culture then emerged.

If this is correct, then culture is not a single evolved trait comparable, say, to the structure of our external ears or our hands. Rather it is a supertrait (standard biological terminology fails us here), the consequence of a number of constituent psychological mechanisms that initially evolved for reasons other than the advantages of culture. But it is even more complicated than this, because, by definition, culture is something that involves a linkage of some kind between individuals such that they form a group, the results of which are to the advantage, including the survival advantage, of the individuals involved in the sharing of that group. And, as just indicated, this supertrait, for the reason that it confers increased biological fitness upon those linked into a group, may then become a force for evolutionary change in the psychological mechanisms of individuals. Such changes within individuals then feed back into the character of the supertrait acting between individuals.

This shifting of focus between individuals and groups is one way of depicting the complexity of culture. There is another. Remember from the last chapter the possibility, indeed the likelihood, that the basic mechanisms of neural plasticity evolved soon after the appearance of nervous systems in multicellular animals. If neuroscience in the coming decades is able to show, as I believe it will, that neural plasticity is the result of a single set of biochemical cellular events acting within a single neural microstructure, this will mean that the manifest differences between different forms of cognition are the result of differences in the ways the basic microstructural units are configured into cognitive macrostructures. These macrostructures in turn are the result of hundreds of millions of years of evolutionary tinkering with, and moulding of, the basic mechanisms of plasticity, and constitute adaptive co-option of function on a grand scale. It really matters not whether we call these different forms of intelligence adaptations or exaptations. It is the complexity that we should marvel at, a complexity that strains the conceptual resources of evolutionary biology.

Let's put it another way using as an example a specific psychological mechanism. This is where the caveat about regarding each form of intelligence as a single adaptation or exaptation must be sounded. Everyone agrees that language is one of the essential components of human culture. But language itself is probably a supertrait that arose out of a number of previously evolved mechanisms, such as working memory, a sensitivity to the temporal order of sensory input, and the capacity for segmenting input into higher-order chunks (to name but three among the many other constituent processes and mechanisms that have been put forward by theorists as the essential components of language). Thought of in this way, the capacity for symbolic language is not a single adaptive mechanism, but emerged out of the interaction between the consequences of a chance alteration in genetic structure and already existing adaptive psychological mechanisms. This may well be true also of other complex cognitive capacities. If that is the case, culture is an adaptive supertrait built upon – emerging from – cognitive mechanisms, some of which themselves might be emergent supertraits, and all of which have themselves evolved as a result of repeated co-opting to function of neural network structures. What we have here is adaptation building upon adaptation building upon adaptation – exaptations of exaptations, if you prefer – on a scale unlike any other encountered by biologists. And that is one of the reasons why culture is such a difficult phenomenon for the natural sciences to explain.

Broadening the picture

In the preface, the writing of this book was likened to the weaving of a tapestry. Some of the threads represent different aspects of biology, including psychology; other threads stand for various concepts and approaches from the social sciences, which, after more than a century of study, can reasonably claim to know something about culture. So far we have been working largely with the biological threads. It is now time to incorporate some social science into the tapestry. We can do this by looking again at those chimpanzees, which show such striking regional variations across a range of behaviours. As correctly argued by the authors of the paper in *Nature* referred to earlier, neither genetic nor ecological variables can explain their observations. What does explain them is learning. The chimpanzees have learned to do certain things by observing one another, the different behavioural traditions being the result of something which is shared between animals confined within isolated social groups. This is not culture as we know it in humans, but it is culture of a kind. I have brought it up again because although it is a highly simplified kind of culture it has an obvious quality, a core quality that most scholars would judge to define, in part at least, this thing called culture. This is the quality that something is shared. Of course, the individuals that make up a social species share many attributes, and there are three sources of such sharing. Most shared attributes are due to the individual animals being members of a single breeding population and the traits that they have in common such as vascular system, dentition or limb form are species-general and the result of a common gene pool and similar environments of development. There is, to be sure, some variation across what is shared, but this is well understood in terms of differences in individual genotype and small differences deriving from the complexities of development. The second source of sharing relates specifically to behaviour in social species that have evolved intelligence. In these species some of the behaviour of individuals may be shared because, in one way (by chance, or because they all have the same forms of intelligence and face the same problems) or another (drawing one another's attention to particular features of the world), they have learned the same things, but learned them individually. The third source of sharing results in what has been observed in chimpanzees and songbirds: the shared behaviours are not only learned but they are learned *from others*, who learned them from others again before that. The transmission of something learned by an act of learning is the core quality that anchors our recognition and understanding of what culture is.

In pluralist and generous mode, then, chimpanzees and songbirds (and perhaps other species out there in the still little-explored but large biological world) are judged to have culture because they are able to learn from others, and what they learn – relatively simple motor acts – has been learned from yet others, who also acquired it by learning. In the case of chimpanzees, there was, of course, an initiator, an animal who invented the act in the first place, but that is beside the point of what the core quality of culture is. Now, if you are one of those who insist on limiting application of the word *culture* to us humans, call it protoculture, because there is no question that what we humans share, and what sets us apart from all other species, goes far beyond that which chimpanzees share. However, it is precisely in the claims as to just what it is that we share that the differences between the various social-science approaches to culture lie.

The social-science study of culture has been overwhelmingly the province of anthropology. Anthropology has existed as a science for at least as long as psychology, and a sketch of the history of the culture concept as formed by anthropologists would take an entire book in itself. What follows in the next few pages, therefore, does not, of course, constitute an exhaustive survey. But for our purposes it will suffice. Because of the necessarily limited nature of this review, a recent book by Adam Kuper, a London-based anthropologist, is strongly recommended. This authoritative yet highly readable survey of the history of the concept of culture is listed at the end of the chapter. (For the rest of this chapter, but not in others, I am going to use a device employed by some anthropologists to distinguish between Culture with an upper-case C, which is whatever one thinks it is that is shared – symbols, values, etc. – and culture with a lower-case c, which is the specific manifestation of Culture in a particular social group.)

An outsider coming fresh to the anthropological literature is struck by several things. The first is the seemingly radically different views taken by anthropologists of what Culture is. Sometimes these differences are so great, apart from conformity to the notion of something shared, that the reader may be forgiven for having a sense of misgiving about the very word itself. This sense of misgiving is clear in Kuper's fine book, and is one of the reasons why I keep saying Culture is different things. However, one must cling to that core quality as defining a specific phenomenon: learning from others what they in turn have learned. This kind of learning does occur and its consequences are real. Culture is not like phlogiston or the ether. It exists. It is, was, and always will be a major characteristic of our species. When those Martians of a thousand thought experiments have returned to Mars and report on what kind of creature Earthlings are, they will say that we share something with each other – and if they have had training in anthropology, the crew will most likely disagree as to what it is that is shared.

The second feature of the anthropological literature that strikes the outsider is that it carries some conceptual baggage which seems not to be helpful, and at times is downright confusing. So the first thing to do before summarizing the way anthropology has dealt with Culture, its key subject, is pick this baggage out and where necessary discard it. One of the things to be discarded is use of the word *Culture* when what is being referred to is the 'possession' of a knowledge of the arts and humanities (and, sometimes, science) by an elite within a society. This is the usage that famously had the Nazi Hermann Goering reaching for his revolver, and a strong connotation of it is mutual loathing between social groups, social classes or even whole countries and continents. The image evoked is of a beer-drinking, karaoke-loving oik being looked down upon by a champagne-swilling opera buff. This was the sense in which Matthew Arnold used the word. And this 'them' and 'us' elitist usage is sometimes expanded to encompass whole national or political movements and equated with 'civilization'. It has not been uncommon for European nations to disparage, say, English culture or to applaud French culture. Whole continents can be drawn into this, with insults being exchanged about European and American culture. Now while such behaviour may be, and often is, part of the content of a culture – of that which is shared – it has absolutely no place in an analytic, explanatory approach to what Culture is. It is simply not science of any kind.

Another issue that often confuses the newcomer is the apparent hairsplitting by anthropologists, especially over distinctions between the cultural and the social, between Culture and society, between social systems and social structures. It is always the case, however, that outsiders cannot see the importance of the minutiae of science. In this instance, 'outsiders' includes most biologists, many of whom have been trampling all over these distinctions for years. It actually is important that words retain their specific meanings in science, and conflating Culture with all things social threatens that thing feared by all social scientists when they see a biologist bearing down on their subject – destruction by oversimplification. At times it is indeed hard to maintain a clear distinction between Culture and various aspects of the social because they are so closely interwoven. However, maintain the distinction we must if we are to get a real hold on Culture. In his masterly review of theories of Culture written in the 1970s, the Australian anthropologist Roger Keesing makes the distinction thus: social systems are 'designs for living in particular environments', while Culture, along with history and the conditions of the environment, is what causes social systems. Methods of subsistence, settlement patterns, even the relationships in which people engage, are all aspects of the social system, which is the embodiment of the interaction between Culture, history and environment. And as outlined in chapter 1, the social and the cultural feed back into one another. The social construction of baaskap, one of the key elements in the culture of apartheid South Africa, led to specific material conditions – the social system of laws, the locations where people lived and worked, what work was done by whom, and all those other things that comprised the social system of the country – which then fed back into the social constructions that characterized that specific culture. For our purposes the distinction is relatively straightforward. *Culture* is reserved for that 'something shared through learning'. Culture, to be sure, gives rise to material objects, such as stone tools or clothes, immaterial objects, such as United Nations resolutions or acts of parliament, and social organizations, like schools and labour unions. In short, Culture has consequences, and those consequences, out there in the world, are commonly referred to as 'social this' and 'social that'. But Culture is not its effects. It is something in itself. The word *social* is also commonly used in the context of close physical proximity, but a social group is not necessarily a cultural group. A herd of wildebeest is not a culture, and neither need be a social group of humans.

This brings us to a third bit of tidying-up. The word *Culture* is sometimes deemed to have such overarching meaning as to have almost none at all. A. R. Radcliffe-Brown, an early leader of British anthropology, referred to Culture as a 'vague abstraction'. For Kuper a concern is that this 'hyper-referential word' becomes problematic when Culture can be something to be both described and explained on the one hand, but also a source of explanation on the other. For example, the use of animals as emblems of kinship groups, i.e, totemism, may be revealed by the ethnography of a group of people (ethnography is the systematic description of a culture, which was the empirical bread-and-butter of classical anthropology). And it can then be used as the explanation for certain kinds of behaviour, such as dietary customs. Now exactly why this is problematic is not clear. After all, it is common in biology, and especially so in the psychological sciences and neurosciences, for something to be both caused and causal. How the brain and psychological mechanisms have come to be what they are and how they work are what neuroscience and psychology are about. These disciplines seek a causal explanation of neural structure and psychological function, and that the brain causes psychological states and behaviour is neither in doubt nor problematical. Behaviour, too, is both an effect, i.e. it is caused, and has effects, i.e. it is causal. This also is neither doubted nor problematical. Well then, there should be no special difficulty in understanding Culture in this regard.

However, one possible spin on, or interpretation of, Kuper's concern is given in an essay by the American anthropologist John W. Bennett, published in a recent issue of the journal *American Anthropology*. Bennett, surveying the history of classical anthropology, argues that 'if you postulate Culture as a major cause of human phenomena, you create an epistemological problem' because you are 'defining a thing by itself. Culture *is* human phenomena; human phenomena, generalized, *are* Culture'. But surely this is, as they say, going over the top. Equating Culture with all human phenomena is simply wrong. Putting aside more basic biological characteristics, such as bipedalism, which have nothing to do with Culture (or cultures), there is a whole raft of behavioural and psychological characteristics which have nothing to do with it. For example, the ability to detect physical causation, often referred to as intuitive physics, is a human phenomenon that is universal and unrelated to Culture. So too is the ability to construct a coherent visual picture out of the impulses flooding into the brain along a million fibres in each optic nerve. There are countless other such examples. This is not to deny the causal force of Culture and the extent to which it permeates human thought and activity. But it is simply undiscriminating to run all human phenomena together as Culture.

Yet later in the same essay, Bennett switches his focus to a narrower epistemological issue that chimes with Kuper's concern. Such is the importance of this to any attempt to build bridges between sciences that a couple of further paragraphs need to be devoted to it. Bennett takes a long historical view of anthropology's dealings with Culture and thinks that the problem of 'assigning causal significance to Culture when at the same time Culture itself is created by the behaviour it is supposed to cause' came to occupy a key role in anthropology's turning-away from Culture in the 1960s and 1970s. In this context he notes how anthropological study of Culture might have been saved from this supposedly fatal stance of facing two ways at once – of Culture being at once its own cause and the explanation of what it is caused by – by the work of those, including the polymathic Gregory Bateson, who might have broken out of this circularity by establishing links with other disciplines, notably psychology. But the anthropologists,

Bennett claims, were too frightened of losing the heart of their subject to another discipline. So they were left paralysed by the mixture of an intractable circularity and the fear of encroachment by other sciences, and by the 1970s had turned to the study of specific areas of social phenomena, ranging from politics, through medicine, to ecology. In effect, argues Bennett, rather than lose their central idea to other scientists, they abandoned it altogether.

Well, one need not judge the historical accuracy of Bennett's argument, though it certainly is the case that for 50 years and more anthropology was dominated by the approach of Franz Boas, the founding father of American cultural anthropology. The Boasian view was that Culture reigns supreme as the cause of what humans are, that biology of any sort is of trivial consequence in the making of humans, and that understanding Culture is the unique aim of anthropology. But where Bennett is absolutely right is in his appreciation of the conceptual power to be gained from linking with neighbouring disciplines. That is the motivation behind this book. But the advantages of interdisciplinary study aren't unique to the examination of Culture. All scientific disciplines gain enormous strength by linking with their neighbours. As noted in chapter 1, the idea of the unity of science is the idea of doing just that. And, of course, if Bateson's example had been followed by others, Culture would not have been lost to other disciplines. What would have happened – and, it is to be hoped, will happen in the future – is there would have been an increased understanding of Culture to the mutual benefit of anthropology and the disciplines to which it had reached out. Nobody would have lost anything. If evolution can help us understand Culture better, evolution's standing as a theory will be strengthened. If psychology can help explain what Culture is, both cognitive science and the social sciences will be strengthened.

Nobody, though, should be left with the impression that all anthropologists suffer from massive and ruinous intellectual provincialism or paranoia. For one thing, anthropology departments in universities often enough have both cultural (sometimes called social) and physical (or biological) anthropologists on their faculties, and the latter are pretty well indistinguishable from other biologists. And Bennett is not the only cultural anthropologist to understand the importance of

crossing disciplinary boundaries. Several schools of cultural anthropology have stood, at least in part, quite explicitly in the domains of
neighbouring disciplines.

One of these is the functionalist school of anthropology, which
views Culture in general, and any culture specifically, as a system of
shared customs and behaviours, each of which has a function in the
workings of that system. Well, this is the approach of functionalists of
any kind but translated into the realm of the social. If the body be
viewed as an integrated system, which, of course, it is, what is the
function of the organs of circulation? Transport, is the answer. If the
cell is an integrated system, what, say, is the role played by mitochon-
dria? The answer is transformer of energy from food into a form that
the cell can use. In like fashion, if the culture of a particular social
group is an integrated system, what is the function of this custom or
that belief? And once you start asking such questions of the culture of
one social group, you are driven to ask the same questions of other
social groups and to compare the answers. In effect, anthropology for
the functionalists was a kind of comparative sociology.

But there is a big 'if' lurking beneath the functionalist approach
that leads to further, and deeper, questions. If you adopt this stance,
what is the source of the integration? The heart functions as a pump
which drives a transport system that serves the whole body. To be
sure, that transport system serves the heart as well, but the heart works
for reasons other than its own existence. This is a banal observation
about the biology of the body, but when the question is asked of
Culture, it is less easy to answer. Just what is it that gives a culture its
integrated shape, and is that shape constant across cultures? Here lie
the beginnings of the notion of Culture as superorganism. And what
is the source of a function, a job that is done, when applied to the
elements of a culture? Just as there are bodily structures that have no
function, one might expect some elements of a culture to have no
function, and for the same reasons perhaps. But most of biology
concerns structures that do have functions. Is the study of Culture the
same? And if it is, does this imply something universal, dare one say
lawlike, about all cultures?

What these questions all raise is the issue of generality – generality
as applied to cultures and generality as applied to Culture. Reading

the literature on Culture gives the strong impression that, prior to the postmodernist era, in which relativism militates against any generalities at all, most cultural anthropologists subscribed, implicitly if not explicitly, to the position that generality does indeed exist. Roughly speaking there are three different ways of thinking about such generality. The first could be labelled a *sui generis* approach. Here generality derives from what Culture is, and what it is is a unique phenomenon that can only be understood within the framework of what makes it unique, which is an integrated system of symbols, beliefs and values – 'collective symbolic discourse' in Kuper's words. The curious notion of Culture as a 'superorganism' arose naturally out of the combination of the isolation of Culture from all else, and the sense of its complex articulation as a coherent whole.

The other two ways of thinking about generality are very different from the *sui generis* approach, but not mutually exclusive of one another. The first, and the second possible source of whatever it is that gives Culture that feel of generality, is that Culture functions to assist biology with what it cannot do on its own. This is the Culture-as-adaptive-system school of thought, and it has taken many different guises, some of them explicitly evolutionary in form. According to this approach, Culture is the instrument of a human nature which is open-ended in design and whose 'completion and modification through cultural learning make[s] human life viable in particular ecological settings', in the words of Keesing. There is a strong materialist and biological flavour to this adaptationist view, which, unsurprisingly, is the school of cultural anthropology most congenial to biologists. Cultures are seen as patterns of behaviour and material objects such as tools, machines and other artefacts. These are not directly genetically caused, but are directly expressed as specific technologies and economies, which are considered to be the properties of a culture that above all others help individuals in it to survive. As Keesing notes, religion, rituals and other symbolic activities are secondary properties of cultures, perhaps even epiphenomena.

Another way of thinking about the adaptationist approach to Culture is to see it as viewing the relationship between Culture and biology in roughly the same terms as chapter 2 of this book sets out the relationship between intelligence and the main evolutionary programme. The only difference is that chapter 2 argues that intelligence evolved because of the inability of evolution alone to sample and gain information about biologically significant events which change at certain rates, while anthropologists of the adaptationist school make no assumptions about how Culture came into existence or exactly what the relationship is between it and things biological. What they do assume is that culture ensures survival in any ecology by providing adaptations that human biology cannot. The unexplored nature of the relationship between an adaptational Culture and a needful biology provides the conceptual space for assuming that at least some of Culture's generality has its source in that needful human biology. We will explore this further in chapter 6.

The third approach to what is general about Culture rests on the basic assumption that Culture somehow reflects, or is a consequence of, the structure of the human mind. In its weak form, this is pretty well universal and unremarkable. Unless one invokes supernatural agency, there is no other possibility. But the assumption does have stronger forms – three to be precise. These are usually labelled structuralism and, rather more vaguely, the schools of anthropology that consider Culture either as some form of ideational or knowledge system, or as a system of signs and symbols. Structuralism in anthropology is overwhelmingly connected with the work of the French anthropologist Claude Lévi-Strauss. For Lévi-Strauss, any culture is a product of two sets of factors. One is the structure of the human mind, which is universal and unvarying. The other is the environment, both physical and social, of a particular social group, including the history of that group. It is the latter – local environment and history – that accounts for cultural differences between groups.

Structuralism has its origins in the early-20th-century work of Ferdinand de Saussure, the French linguist who developed the idea that language is to be understood as a structure, a closed system of elements with rules for the transformation of those elements, rules that are linked to the overall structure of the system. Jean Piaget, the Swiss psychologist, described structuralism as the study of any 'system closed under transformation', in a book which readers wanting to pursue structuralism in depth should consult. There is no easy description of this highly abstract set of ideas beyond the general notion of a rule-based set of transformations within a closed system, because, following Lévi-Strauss's introduction of structuralism into anthropological theory, it was adopted in whole or in part, and used as an analytical device, by people working in a range of disciplines, including mathematics, psychoanalysis, cognitive psychology, linguistics, literary theory and political theory. Suffice it to say here that structuralism sought the rules of symbolism in cultures on the assumption that they were linked to the invariant structure of the human mind itself.

Given the basic premise of a universally structured mind, which would include knowledge structures, it is, of course, arbitrary to distinguish structuralism from what Keesing calls ideational theories of Culture. From this point of view, Culture is knowledge. In the words of one of the most prominent exponents of the ideational school, Ward Goodenough: 'a society's culture consists of whatever it is one has to know or believe in order to operate in a manner acceptable to its members. Culture is not a material phenomenon; it does not consist of things, people, behavior or emotions . . . It is the form of things that people have in mind, their models for perceiving, relating, and otherwise interpreting them.' While specifically eschewing material artefacts as Culture, Goodenough's otherwise excellent definition is somewhat spoiled by the assumption, probably not really meant, that knowledge is not material. But what people know, without going into the philosophical technicalities of knowledge as justified true belief, is what they have in their memories and other related psychological mechanisms and processes. And there are no psychologists who are dualists, at least not professionally. Memory, thought, planning and values are states of neural network structures.

Unlike the abstract, highly theoretical and, to many, hand-waving and questionable approach of structuralists to Culture, ideational theorists like Goodenough are exceedingly psychology-friendly. States of knowledge or belief are presented without any preconceptual loading as to how to understand what they are. Knowledge and belief, surely, are whatever the current best psychological theories tell us they are. And this is an approach firmly rooted in ideation, with the specific disavowal of either material objects or behaviour as Culture. To a psychologist, this is the approach that seems most likely to build bridges to a scientific psychology, and it is the approach to which we will return when we come to consider possible psychological mechanisms of Culture. It was exactly this marriage, an ideational stance on Culture as bride and psychological mechanism as groom (or perhaps the other way around), that Goodenough himself pursued over a long period of time. Keesing also perceived the crucial importance of ideation, of the mind, for understanding how meanings, beliefs and knowledge are shared and cashed out into social systems.

A related and prominent anthropological school treats culture as symbols and signs, as semiotic systems, or as texts and discourses, to borrow from some of their jargon. However, while the ideational school could be described as pursuing cognitive anthropology, the same could not be said for the seemingly related symbolists. To the outsider this is odd, because signs, symbols and discourse must stand on ideational and cognitive foundations. But while cognitive anthropology almost begs for a bridge-building exercise with psychology and neuroscience, most of the symbolists would find this intolerable. The members of this school are amongst the most Diltheyan (see chapter 1) of the 20th-century anthropological schools, not counting the postmodernists. Symbolists like Clifford Geertz talk about the necessity of *interpreting* symbols, symbolic objects and events. These are people who think of Culture *as* semiotics, not just to be interpreted or analysed *through* semiotics. Culture is shared symbols and meanings, but existing at a level separate from everything else. Keesing, quite rightly, is sceptical. 'Without informing our models of cultures with deepening knowledge of the structures and processes of mind, our cultural analyses may turn out to be mere literary exercises.' Amen.

Thus it is that in their different ways the 20th-century schools of anthropology sought to identify what is shared by cultures, what it is that makes cultures coherent, and what is general to all cultures and thus defines the essence of Culture. None of the approaches is exclusive of the others. It is perfectly reasonable to study the ways in which Culture, and specific cultures, have advanced human survival, which is what the adaptationist school has done, while others analyse Culture, and indeed the same specific cultures, in terms of their ideational, structural or symbolic qualities. So, insofar as a method of analysis and habitual modes of thought emphasize different aspects of the phenomenon, Culture is not some single thing. A gesture common to a group of chimpanzees is certainly different from the symbolic system of a particular human social group. And the totemic symbols of one culture are different from the chemical symbols of another. It would be foolish to argue otherwise. Yet in each case both things do have in common that core quality of sharing, the source of which is the transmission by learning of something that has been learned. Whatever methodological or theoretical stance is adopted, this is the central feature of Culture.

The trouble with 'levels'

Keesing's warning that some forms of cultural analyses might turn out to be mere literary exercises had its source in a perceptible tendency towards intellectual isolationism on the part of some social scientists. When symbolists claim that Culture exists at a level different from all else, they are using the word *level* to invoke the separateness of their discipline, an unbridgeable gap between their subject matter and everything else. It is a statement of 'hands off' to all and sundry. Culture, we are told, is not in minds, it is between minds, or it transcends minds. Well, this is Culture as magic, and it is without meaning. The destruction by simplification of which biologists are accused is countered by a destruction through mystification. However one chooses to think of Culture, as a means for adaptation, as part-cause of things social, as a system of signs and meanings, as shared knowledge, values and beliefs, or even as simple shared acts, it must have entitativeness. By this I mean Culture must have spatiotemporal characteristics, it must be of this world, and it must make contact of some kind with the other things in this world. Exactly how this occurs may prove very difficult to understand, and may parallel the mind–body problem of psychology. The relationship between the mind and the body is an old philosophical problem. Many solutions have been proposed, but none has yet proved acceptable to everyone. But all scientists, in their science, are materialists and physicalists. Everything is chemistry. No scientist believes that the stuff of the mind is non-material and different from the stuff of the body. Dualism is simply not an option. Well, in like fashion, no scientist should think of culture as having some kind of *élan vital* of its own, of there being culture stuff that is removed from and different from mind and body stuff. As will be discussed in the final chapter, there is indeed a seemingly magical quality to Culture. But Culture itself is not magic.

Ah well, some may reply, to claim that culture is an 'emergent' phenomenon, as the title of this chapter does, is an admission of some degree, if only the tiniest, of the non-material. But this is simply wrong. Emergence does not mean a coalescence out of nothing or into nothing. Emergence is the appearance of a property or phenomenon which is inherent in some constituent processes or entities and which is expressed when those constituent elements come together. About this there is nothing magical or mysterious. The emergent phenomenon has direct causal links to those constituents. It is the product of those constituents. The central assumption of this book is that Culture rests on cognitive and other psychological foundations. These are the constituents from which Culture emerges. Thus Culture is *in* minds and brains. It also finds embodiment in physical structures, social organizations, rules and laws. And these feed back into our minds and brains. Complex Culture is, but non-material it is not. It is certainly the case that the causal power of Culture may have a queer feel about it – even the sensible Keesing refers to the 'magic of shared symbols'. But abstracting Culture to a different level, without saying what that level is and how it connects with other things, gives it an ethereal and other-worldly quality. It takes it out of the realm of science. It certainly takes it out of the realm of natural science.

Level can be a puzzling word when it is used without being tied down. It occupies more than two pages of the *Oxford English Dictionary*. There are two, related, general meanings. The one concerns elevation of the horizontal plane, as in height. The other refers to position or status on some scale. When a symbolist claims that Culture exists at a level different from all else, it seems the word has the former meaning. Some kind of difference of elevation is being invoked. But what dimensions are we talking about here? To be fair, both biologists and social scientists frequently invoke the notion of levels. It is a commonplace to see reference to 'the level of the cell' or 'the ecological level'. Partly the word is being used to provide a focus of attention – 'the level of the cell' focuses analysis on that restricted realm. But equally often there is an undoubted, if vague, appeal to some kind of structural or functional relationship – with imagery that is directly spatial, the cellular level is different from the level of the organism or of ecology. Well, such difference may be characterized in a number of ways, but the imagery and the intention of the conceptualization *always* imply connections between levels. This is the essence of any conceptual synthesis in science. We make better sense of the cell, the whole organism and ecology by understanding the connections between them, rather than understanding each in isolation. This must hold when one is attempting the ambitious synthesis of the social and biological sciences. It is all about making connections. It is all about creating a defensible scheme or structure in which 'levels' are related to one another.

A solution to the levels problem

Hierarchy is one of the conceptions most commonly invoked in pursuit of a solution to the problem of structural complexity in biology. For some theorists, hierarchical structure is a requirement in the definition and diagnosis of life. Herbert Simon wrote a paper, listed at the end of this chapter, with the brilliant metaphorical title 'The Architecture of Complexity', in which he argued that hierarchy 'is one of the central structural schemes that the architect of complexity uses'. He maintained that without hierarchical structure, the complexity of biological systems could not have evolved. In the most general terms, a hierarchy is a form of ordering that is partly dependent upon scale along certain dimensions, such as energy levels, frequency or size. There are two distinct types of hierarchy. One is the structural hierarchy and the other a hierarchy of control. Failure to recognize the difference has often led to confusion.

The principal feature of a structural hierarchy is containment, sometimes referred to as the Chinese puzzle or Russian doll characteristic. Take something, open it up, and inside there are other entities. Open these up, and inside are yet other things, which if opened in turn will reveal that they too contain entities. You keep doing this until you find something that cannot be opened further, at which point you are at the fundamental level of that structural hierarchy. All living things are structural hierarchies, this being most obvious with multicellular animals. These contain organ systems, which are made up of tissues, which are composed of cells, which contain organelles, which are made up of macromolecular structures. And, of course, going the other way, individual organisms may cluster to form groups or communities, which in turn are constituents of breeding populations, which make up species, aggregates of which comprise larger ecological units. Such structural hierarchies scale by size and the strength of the forces that bind the constituent elements into coherent wholes, both at every level and between levels.

Simon famously told the story of Hora and Tempus, two watch-makers of fine, and complicated, instruments. Their watches consisted of some 1000 parts, and their telephones kept ringing with new orders. Hora prospered, but poor Tempus's business failed. The reason for this was that Tempus had never discovered how to structure his watches as subassemblies of parts, which was what Hora had done. So whenever Tempus's phone rang with a new order and he had to put down the watch he was working on, the watch fell apart, and when he picked it up after hanging up, he had to start all over again from the beginning. Poor Tempus. He seldom got to build a complete watch, so developed a terrible reputation for being unable to deliver. Hora, in contrast, had discovered structural hierarchies, and when he put down an unfinished watch to answer the phone, it didn't completely fall apart. He constructed his watches by building ten-part subunits, then put ten of these together to form yet larger units, ten of which he combined to make up a complete watch. So while Hora did lose some time when he put a watch down, it never fell apart into a state of total disorder like one of Tempus's. Simon's parable of the watchmakers illustrates the stability of complex structures when they are hierarchically ordered, hence the reason why complexity might have evolved the form that it did.

Hierarchies of classification – for example, of living forms – conform in some respects to structural hierarchies. Species are 'contained' within genera, which cluster to form families, orders, classes and phyla. Such hierarchies, however, are all in the mind. We create them as a mirror of the world in order to make sense of it, but they are not really structural hierarchies. Insofar as they are depictions of causal events in the world, they provide us with a bridge to that other form of hierarchy, based on control.

Control hierarchies are formed by dependencies, causal chains free of containment. The chief executive of a company is at the pinnacle of a control hierarchy. He or she exerts control over the other employees in the organization. While it could be a very flat hierarchy, with all other employees being equal in authority, this would be unusual in a large company. Usually there are further levels of control, such as ordinary executives, heads of department, heads of section, and others of lesser rank. There is no physical containment. The chief executive is not made up of the other people in the company, and the scaling is measured by degree of authority or control. The chief executive has greater power to cause things to happen than have the executives, who in turn have greater authority than departmental heads. One of the characteristics of control hierarchies, which varies with the degree of control vested at each level, is that they are dynamic, with dependency, be it control, flows of information or both, moving fluidly in all directions. Under the guidance of the board of a company, the chief executive makes decisions that are conveyed down the hierarchy to other members of the company, while even those at the lowest levels of the organization can report events and problems to their superiors so that they feed upwards to affect decisions made subsequently by higher authority. Indeed, causal power itself is not fixed in complex control hierarchies but may shift with changing circumstances from one point in the system to another.

Social systems are often control hierarchies. So, too, are complex organ systems in our bodies, such as the immune and nervous systems. In the brain information flows along sensory pathways to neural networks, which process the input and pass it on for further processing and integration. Feedback and feedforward loops operate to adjust the functioning of the whole system. Individual cells also operate as complex control systems, with intracellular chemical events turning the expression of genes on or off, which in turn leads to the production of complex molecules and molecular clusters. The development of a single fertilized egg into an exquisitely structured complete organism with billions of cells is a cascade of control hierarchies.

The causal chains of control hierarchies spread far and wide in living systems. The relationship between biological evolution and individual intelligence described in the previous chapter conforms to a control hierarchy. Evolution is one level of this hierarchy and intelligence another. Evolution is the more fundamental level, or the level of greater authority, for it is evolution that caused intelligence to come into existence. In the jargon of hierarchy theory, intelligence is nested under evolution, in that what it does, what it is intelligent about, are the things that the main evolutionary programme tells it to be intelligent about. Those, remember, are the very things that evolution itself cannot know about in detail because the rate at which evolution occurs is too slow relative to the frequency with which some circumstances in the world change. The constraints on intelligence, the different forms it takes in different cognitive modules within an animal, as well as the differences in its form from one species to another, are the consequences of this relationship. However, as in all control hierarchies, this is a dynamic system. Intelligence can have evolutionary consequences. When intelligence enters into decisions about where an animal lives, what it eats or with whom it mates, these consequences can be potent.

In a previous account of how humans, and other intelligent animals, evolved the ability to acquire knowledge of their worlds, I used the word *heuristic* to describe the inventive, creative processes of both evolution and intelligence. In everyday language, heuristics are things – devices – that lead to discovery and invention. Since that is what both evolution and intelligence do, each, as a cluster of processes and mechanisms occupying different levels of a control hierarchy, were respectively referred to as primary and secondary heuristics.

It is within this framework that Culture can be located. Culture emerges as a supertrait from certain essential cognitive components. The causes of culture *are* those cognitive components. Culture, then, is a further level within the control hierarchy. Just as intelligence is nested under evolution, so Culture is nested under intelligence. Culture, the tertiary heuristic, is constrained by intelligence. It is not a level that floats free, rather it is connected to the things that cause it to be. Again, the system is a dynamic one. Culture has consequences for both intelligence and evolution.

Within this formulation, the secondary heuristic, intelligence, is what is making the connections – connections between biological evolution on the one hand and Culture on the other. The characteristics of intelligence so belaboured in this and the previous chapter have been belaboured precisely because they are the characteristics that tie together the biological and social sciences. They are the negation of genetic reductionism, countered by the structure of intelligence, which points to the constraints that lie in its own evolution, and how certain inherent properties of its constituents in humans allow for the emergence of culture in the form of something, or many things, shared. To continue with the metaphor of marriage, these properties are the bonds of the marriage of the social and natural sciences. If there is meaning in symbols, as the symbolists claim, it is the meaning in people's heads that is shared, and it is intelligence that does the sharing.

There is one final point to be made before taking hold again of the biological threads of our tapestry in the next chapter. When the full story of Culture is eventually known, it will involve a hierarchy much more complicated than that just told. Development was described earlier as a cascade of control hierarchies, and these must be inserted between evolution and intelligence. And standing between intelligence and Culture is another aspect of development, which is enculturation – that is, the way intelligence develops within Culture. Imagine a kind of conceptual club sandwich with development occupying the space on either side of intelligence, and biological evolution and Culture forming the outer layers. This, however, is a dynamic club sandwich, which folds back on itself so that each layer interacts causally with all the other layers.

We do not yet know enough about either the development of the brain and its attendant psychological mechanisms or how the brain is related in any real detail to those psychological mechanisms. And, while it is a matter of increasing debate, we presently know little of how to stitch together evolution and development. With time, our knowledge of these matters will improve, but in the meantime our architecture of complexity will remain only partial and incomplete. Ignorance, though, should not stand in the way of attempts to build a partial synthesis. We can work with just three levels provided we keep in mind that the scheme is incomplete – incomplete, but not oversimplified in respect of the social sciences.

Suggested Readings

Keesing, R. M. (1974) 'Theories of Culture.' *Annual Review of Anthropology*, vol. 3, 73–97. (A fine review of classical culture theory.)

Kuper, A. (1999) *Culture: The Anthropologists' Account*. Cambridge, Mass., Harvard University Press. (A readable and cultured account of anthropology's struggles with Culture.)

Piaget, J. (1971) *Structuralism*. London, Routledge and Kegan Paul. (A brief, if difficult, account of a difficult school of thought.)

Simon, H. A. (1962) 'The architecture of complexity.' *Proceedings of the American Philosophical Society*, vol. 106, 467–82. (A classical account of hierarchy theory.)

Whiten, A., et al. (1999) 'Cultures in chimpanzees.' *Nature*, vol. 399, 682–5.

4

Naturalizing Culture the Process Way

In a famous where-we-are-now statement of the early 1950s, Alfred Kroeber, one of the leaders of cultural anthropology, made some very strange comments about the relationship of his subject with the natural sciences. He simultaneously affirmed the Boasian premises that culture is the principal determinant of human nature and the central subject-matter of anthropology, and acknowledged, through some curious phrasing, that culture must also be seen, somehow, as falling within the remit of the sciences. This is part of what he wrote:

I submit that man as a set of social phenomena, including his culture in all its aspects – along with values – not only is *in* nature, but is *wholly* part of nature. It is evidently going to be somebody's business to deal scientifically with these human phenomena; to work at more than aesthetic comprehension of them. Such a comprehension would be intellectual, aiming at an intelligible concord with reality, resting on both specific evidence and on a broad coherent theory. [Emphases in the original.]

It is hard to interpret this other than as the leading cultural anthropologist of the day asserting that anthropology was not doing science – at any rate, not doing science as it is done within the natural sciences – and that it was time a scientific study of culture was put in place. These comments appear the more startling when one realizes they coincided in time with the onset of some amazing advances in biology, notably molecular biology. Kroeber seemed to depict anthropology as the plodding non-scientific study of culture, so the contrast with biological disciplines could not have been greater. From almost any perspective, this was a strange thing to do. Ethnographies had been gathered with meticulous care for decades, and many cultural anthropologists, including stars such as Margaret Mead and Ruth Benedict, had for long dwelt on the links between culture and human psychology – between the 'parts', or individual minds, and the whole, i.e. culture. Well, as pointed out in chapter 1, much hinges on where one thinks one discipline ends and another begins, and just what one considers the links between the natural and social sciences to be. It is, to repeat the point, all about making connections. Perhaps what Kroeber's statement meant was that at the time he saw no connections of any substance. Still, it was a harsh judgement to make of his own subject.

There are, in fact, two general kinds of connections to be made. One, which is the focus of this chapter, is by way of a cluster of notions relating to process. The other concerns mechanism, and will be dealt with in the next chapter. These are not exclusive approaches. It is almost impossible to discuss one without involving the other, and just how they are integrated is an issue of real importance to a natural science of culture. First, we must turn briefly to the distinction between process and mechanism itself.

A process is a sequence of events that occurs in time and leads to an outcome, a result or end-state of some form. It is a means of proceeding or doing, involving change. Growth is a process. So, too, is thinking. Evolution is also a process. As Daniel Dennett points out in *Darwin's Dangerous Idea*, because Darwin had not the slightest idea as to what mechanisms might underlie evolution, he presented his theory almost wholly in process terms. When a process can be pursued without regard to mechanism, as an entirely abstract set of procedures, it becomes cast in terms of 'substrate neutrality', in Dennett's nice phrase. And when it can be broken down into a sequence of component subprocesses which, with necessary repetition through the cycle of such subprocesses, reaches some end-state, such a substrate-neutral process is called an algorithm. Most common arithmetical procedures are performed as algorithms, and it matters not, as Dennett notes, whether an algorithm is run 'long hand', using pen or paper, or through laborious 'mental arithmetic' with the products of the subprocesses recorded entirely in neural networks – the sequence of operations and the result are the same. 'The power of the procedure is due to its *logical* structure, not the causal powers of the materials used in the instantiation, just so long as those causal powers permit the prescribed steps to be followed exactly.'

Mechanisms, on the other hand, are structures of mutually adapted parts. They are things, moulded entities, they have substance. In principle, you can see, touch and even eat them. You can't touch or eat a process, but all processes work by being embodied in mechanisms. Yes, growth is a process, but the process, as yet little understood, is played out by molecules interacting within the environments of cells, which divide and further interact with other cells, which trigger yet more intracellular activity. The physicist David Bohm once declared that 'all is process' and that 'Nothing is permanent. Change is what is eternal', thus emphasizing the dynamic abstractedness of process, and lamented that 'just when physics is (thus) moving away from mechanism, biology and psychology are moving closer to it'. From this point of view, mechanisms (and anything else that has substance and hence can be touched and eaten) are processes that are changing very slowly. Well, Bohm's is an elegant argument. But it could also be said to be overplaying the notion of the unity of science. As pointed out in chapter 1, we are not all physicists – for good reasons. The essence of flying, echo-locating bats lies, among other things, in the mechanisms – the structures – that allow bats to fly and to navigate using sound. These are anatomical features and physiological mechanisms, as well as the genetic structures and their transformation by development into these phenotypic traits. But, of course, bats in general are the product of a process – evolution; and, as just pointed out, each individual bat is the outcome of the processes of development.

Neither processes nor mechanisms have a monopoly over causal explanation in biology (in which I include psychology). This brings us to a crucial dissimilarity between the biological and social sciences that was not mentioned in our earlier discussion of the differences between them. This is that the metaphysics – by which I mean the basic conceptual tool-kit, including often unstated assumptions – of the social sciences does not include mechanisms. True, some social scientists, such as economists (and some psychologists too, especially in the bad old days of behaviourism), use the word *mechanism* to describe a rule of interaction. Thus market forces are based on 'mechanisms' of supply and demand, but these are not mechanisms as biologists understand them – indeed, they are much closer to substrate-neutral processes. You cannot eat or touch the mechanisms of the market. To do that they would have to be embodied in the minds of people, at which point, of course, you would have moved from one discipline to another, which is a wholly good thing to do but one with which some social scientists are uncomfortable. Put in other terms, getting at the mechanisms that underlie the processes of culture is one way – a powerful way – of making defensible connections between the social and biological sciences. That, though, is for the next chapter. First one must get a handle on just what those processes are. For the rest of this chapter, I want to concentrate on the ways in which biologists have thought, and are thinking, about the processes of culture. Bearing in mind that in the biology of today it is hard to talk about processes without any mention of mechanisms, this is interesting in itself because most biologists, by training and as a result of the history of their subject, are, as Bohm noted, biased towards mechanism. So biologists who talk only about process are, by definition, unusual and interesting. They are theorists in direct line of intellectual descent from Darwin.

Thinking about culture in terms of processes is forced by another feature of human beings – one, indeed, that has been the unstated driver behind the social sciences' stance which so emphasizes processes and underplays mechanisms. This is the way in which people in different circumstances often behave in seemingly similar ways but for different reasons – often manifestly different reasons – which reside in those different circumstances. Sometimes this is an illusion arising from overly crude descriptions of what people do and why they behave as they do. But it also arises out of the complex causal architecture that rules what we do, and the abstract nature of accounts of process gives, at least initially, a better handle on understanding that complexity than does the nitty-gritty, fine-grained detail of mechanism. So, before considering different process approaches to culture, let's first consider an all-too-common example of people apparently doing the same thing but for very different reasons.

The puzzle of war

In February 1916, the German army attacked the French at Verdun, where a fortress stood on a salient in the French line in northern France. Since the taking of Verdun was of little value in itself, the ensuing six-month battle was waged, one must assume, in order to inflict maximum casualties on 'the enemy' and to sap the morale of enemy troops. Over half a million men were killed and as many again were wounded. In order to reduce the pressure on the French, their British allies attacked the German line on the Somme in early July. On the first day of the British offensive, 20,000 men walked to their deaths and a further 40,000 were wounded. The total number of killed and wounded on the Somme and at Verdun approached two million. Careful study of the state of mind of the men who participated in these and other battles reveals a small number who revelled in the slaughter. The great majority, however, were frightened on first arriving in the trenches (as at any battle front); most then subsided into a state of numbed stupefaction. Casualty rates on the Western Front were so high on both sides that soldiers could arrive at a rough prediction of how long they were likely to stay alive. There are few reports of any feelings of anger or aggression towards the enemy. Often anger and aggression among the troops were directed at their own commanders. Claims that men fight wars because war is an extension of individual aggression by which males can gain status, especially later-born males as opposed to first-born, who are privileged by customs of inheritance, or because war is a means of gaining access to women, may be correct in a small number of well-documented cases of warfare. Usually, however – and historical record must be testament to this – people, men as well as women, go to war for no discernible biological advantage to themselves. Pillage and rape are often the consequences of war, but not usually the cause.

As Barbara Ehrenreich points out in her survey of theories of war, people have gone to near-suicidal lengths to avoid fighting in wars. 'Men have fled their homelands, served lengthy prison terms, hacked off limbs, shot off feet or index fingers, feigned illness or insanity, or, if they could afford to, paid surrogates to fight in their stead.' She quotes a 19th-century Egyptian governor who noted of peasant re-cruits that 'some draw their teeth, some blind themselves, and others maim themselves' as a means of gaining release from military service. Nineteenth-century Prussian military manuals forbade camping near forests because to do so would result in troops disappearing into the trees. And, when they arrived on the battlefield, as Lieutenant Colonel Dave Grossman documents in his book *On Killing*, careful study by military historians and observers of major wars of the last couple of centuries records that, prior to the special training methods introduced after the Second World War, the majority of soldiers would not fire their weapons with the intention of killing the enemy – either they would not fire at all, or they aimed their fire into the ground or air or fired 'blind'. Had this reluctance to kill not been widespread, the casualties of war would, of course, have been even greater than they were over the last several centuries.

Here, then, is a paradox, a deep mystery. Wars have been waged throughout recorded history. If there is any trend at all, it is an increase in the frequency of war. The International Committee of the Red Cross reports the occurrence of dozens of wars in the half century following the Second World War. Yet most people participate in war with extreme reluctance and at great and palpable risk to themselves. The personal rewards are usually negligible. What, then, causes wars and why do people fight in them?

Documented exceptions do not prove the general rule when it comes to understanding why people do what they do. Sociopathy in various forms undoubtedly exists in some people, but it is not a common characteristic in any known society. It is an aberrant condition. And, yes, there are men – and the historical record shows it usually *is* men – who kill enthusiastically when social sanctions against killing are lifted and it becomes not just permissible but desirable. But, again, all the evidence points to this being a minority characteristic. (Both Ehrenreich and Grossman are listed at the end of this chapter for those sceptical of this claim.) Furthermore, warfare is not the only example of behaviour that appears to have negative biological advantages for individuals. Celibate priesthoods are another example, as is the killing of children by parents in some societies.

If the study of human behaviour and psychology has taught us anything, it is that people seldom do things for just one reason. Human behaviour, as the psychologist George Mandler insists, is overdetermined, or multiply caused. This is surely correct. So, as to the causes of war and why people fight in them, there are, broadly speaking, two explanations. And they are not mutually exclusive. The first is that people do what they do because of the advantages bestowed on them as individuals, because of gains in their own biological fitness, or because what they do helps them to propagate their genes by gaining advantage for close genetic relatives. The anthropologist Napoleon Chagnon insists that the pattern of warfare among the Yanomamo people of southern Venezuela and northern Brazil can be explained in terms of increases in what is technically called the inclusive fitness of individuals – that is, in terms of facilitating the propagation of each individual's genes. In order to explain this, we need to take a short conceptual diversion.

Darwin was concerned in *The Origin of Species* with the problem posed for his theory by sterile social insect castes. Since the insects are unable to have offspring, who gains from their multitude of structural and behavioural adaptations? Despite individual fitness being the focus of almost all his arguments, what Darwin concluded in this case was that it was the 'community' of which the insects were a part that benefited. This shift in emphasis occasioned no great comment at the time, but was to do so through the writings of others almost a century later. This is because sterile insect castes are an example of a wider phenomenon that became increasingly well documented as the study of animal behaviour advanced. More and more cases were reliably reported, across a wide range of animal types belonging to several phyla, of what came to be called altruistic behaviour. Altruistic behaviour is behaviour by one animal that increases the fitness of another animal at the possible expense of the one doing the behaving. An oft-cited example is the broken-wing display by some birds to draw the attention of a predator away from their nestlings. Indeed, all parental behaviours fall into this category, as do a host of other kinds of 'assisting' or 'caring' behaviour. Another example is fierce defence of the nest by just those sterile insects that Darwin pondered. One explanation of this behaviour was the same as that offered by Darwin, and was labelled 'group selection'. Such behaviour was for the good of the group, or perhaps the good of the species, and is truly unselfish. However, strong arguments were made that such behaviour could not evolve and be stable over any significant period of time because of the likelihood of mutant 'selfish' individuals arising, who, free of the constraints of acting for the good of others, would cheat and freeload on the benefits of the unselfishness of others, and who could concentrate all their energies on behaving in ways that suited only their own individual reproductive success. Because such mutant individuals would have larger numbers of offspring than unselfish animals, the latter, assuming cheating and selfishness were heritable traits, would inevitably fall behind in terms of reproductive success and eventually be driven to extinction.

Furthermore, careful observation of altruistic behaviour indicated that the recipients of helpful activity were often close genetic kin. In the early 1960s, the English biologist William Hamilton formalized this thinking in mathematical terms. Basically, what he emphasized was that genes can be propagated in two ways, not just one. The first is reproduction by an organism itself – that is, individual fitness. The second, often referred to as inclusive fitness, is reproduction of individuals who share some, perhaps many, of that organism's genes – that is, reproduction of genetic kin. Subsequently, Richard Dawkins popularized what came to be known as selfish-gene theory in his book *The Selfish Gene,* as did the Harvard-based biologist E. O. Wilson in *Sociobiology: The New Synthesis.* What inclusive fitness does is widen the explanatory base for behaviour while simultaneously focusing sharply on what the American evolutionist G. C. Williams termed the necessity for propagation of 'the dependent germ plasm'. In other words, it is the continuance of genes that is the fundamental driving force of evolution.

By the late 1970s group-selectionist thinking had taken a fearful battering. As we will see in chapter 6, though, it had not been killed off. Persistence, and no small degree of courage, on the part of a small number of biologists would keep it alive and see it rise again. For the time being, however, gene-centred explanations became almost overwhelmingly dominant. Had selfish-gene theory, or sociobiology as it became widely known, been used only to explain non-human animal behaviour, it would have occasioned merely mild scholarly dissent from some corners of biology because of its overemphasis of the role and importance of genes. Acceptance otherwise would have been widespread among behavioural biologists – as, indeed, it was, to the point where, by the late 1970s, selfish-gene theory had become one of the dominant themes in the study of animal behaviour. But it didn't stop there. It was inevitable that so seemingly powerful an explanation of what animals do should be applied to humans as an explanation of what they do. Human sociobiology led to one of the more vigorous and bitter science rows of recent decades. Labelled a new form of social Darwinism, it was most roundly denounced by social scientists (and by not a few evolutionists and philosophers as well), who saw in it what they most loath and fear – a genetic reductionist explanation of complex human behaviour. As discussed in chapter 2, this was always an unfounded fear in any species that is intelligent. Nonetheless, selfish-gene theory does have at its core the belief that replicating DNA is the process that drives all processes, including evolution. For that reason it was considered by most social scientists, incongruously, to be both vacuous and dangerous.

It was within this context, then, that Chagnon believed Yanomamo warfare could be explained. The Yanomamo really do fight over women, with, so Chagnon asserted, the victors increasing their inclusive fitness. Indeed, Yanomamo culture at large was to be understood, at least to some extent, in terms of individual inclusive fitness. Well, perhaps. Chagnon's data and interpretations have been seriously questioned in some quarters, although the anthropological record contains many observations of other cultures in the Americas, Africa and Asia in which aggression in warfare is rewarded by greater access to women. In this respect the Yanomamo people are not unique. Yet, what selfish-gene theory was being used to explain went far beyond Yanomamo warfare. Mating strategies and mate choice, parental behaviour, social interactions of all kinds, who leaves what to whom in their wills – all became grist to the sociobiological mill. The problem is that some people lost sight of the important truism mentioned a few pages earlier: what humans do is almost never, perhaps never ever, accountable for in terms of a single cause. If selfish-gene theory can be properly applied to animals, there is actually no reason why it might not also explain some aspects of human behaviour. However, the influence of selfish genes in humans may be obscured, and in some circumstances completely drowned out, by other causes of behaviour. Furthermore, wars waged by frightened and stupefied people, reluctant to use their weapons, from quite different cultures, do not seem to be explicable within the same framework as that used by Chagnon to understand the Yanomamo at war. Something else is exerting causal force here, and that something else, of course, is intelligence expressed as culture. The men who fought on the Western Front were taking part in a war caused by social constructions such as national pride, international influence and financial power; and they were more often frightened than aggressive because they had been enculturated into relatively peaceable civil societies in which aggressive behaviour was frowned upon and controlled. Selfish genes seem not to come into it at all, or at any rate not as directly as Chagnon claimed for the Yanomamo. Neither does access to women. But genes certainly do, because genes are among the wellspring of the causes that make us what we are, even if they often act only indirectly via the processes of individual development and the long series of social interactions by which each one of us is inducted into our own particular culture. So, like culture itself, warfare is not a single phenomenon. There are different kinds of war and different ways of behaving in war because there are different causes of war.

How, then, to reconcile such differences and make connections between such multiple causes? Well, all attempts to do this have been within the general framework of thinking about processes. One of these attempts, the architecture of heuristics of the previous chapter, is a form of universal Darwinism. The primary heuristic of biological evolution has given rise to individual intelligence, and from that intelligence, which locks into the intelligence of others, culture emerges. That is a very general conception. There are also more specific and formal ways of considering how these complex and multiple causes are related to one another. It is to these accounts of culture as process that we can now turn.

Universal Darwinism

It is no coincidence, nor the result of whimsy on my part, that the brief account of evolutionary theory given in chapter 1 was framed initially in terms of order, change and diversity. These are process words, and since evolution is a process, such words come naturally in this context. The essence of evolutionary theory is that living systems are transformed in time. Species are not fixed for eternity. Neither, of course, are individual organisms, which change ceaselessly through chemical exchange with their environments, and through development and ageing. Bohm was right: only change is eternal.

Now, Darwin formulated his theory of evolution in terms of slight differences in fitness between individual members of breeding populations, such fitness effects – advantages and disadvantages – being translated into differences in numbers of offspring bearing the characteristics of their parents and hence with the same differences in fitness. Darwin did not understand the source of variation, nor did he know how it was that the traits of parents and their offspring correlated with one another. These are all matters of mechanism. What Darwin was working with was ideas about processes, specifically the process whereby microevolutionary events of adaptation and fitness drive macroevolutionary speciation events. Later developments in the theory, such as the creation of the modern synthesis, either resulted from the embodiment of the process in mechanisms and hence greatly strengthened the explanation of evolutionary change, or added to the process account, as in notions of complexity and self-organization, or changed the linkage between component processes, which is what punctuated-equilibrium theory does by decoupling microevolution from macroevolution.

Because Darwin's original account was process based – and necessarily so, because, despite his meticulous attention to evidence, Darwin never had any mechanisms – it is not surprising that both Darwin and others soon began to apply his mechanism-independent, substrate-neutral theory of transformation in realms other than speciation. It is also not surprising that it was Darwin's 'bulldog', T. H. Huxley, who first considered extending the scope of the theory in this way. Huxley, when first told of the theory, is reputed to have been so impressed with its elegant simplicity and manifest importance that he expressed amazement that no one had thought of it before. He, more than most of Darwin's contemporaries, understood the power of a process-based theory that need be tied to no specific set of mechanisms. About a decade after the publication of *The Origin of Species*, he played with the idea of taking the process of evolution, which for Darwin operated between individuals, and placing it inside single organisms, 'between the molecules', to explain individual development. He pondered the possibility that competition to exist and multiply within organisms occurred, and that 'the organism as a whole is as much a product of the molecules which are victorious as the Fauna and Flora of a country are the product of the glorious organic beings in it'. Oddly, Darwin did not approve. In a letter to Huxley he declared he could not follow him in his 'idea about natural selection amongst the molecules'. It isn't clear whether it was the molecules that bothered him or the shift of the processes into the organism. What could not have perplexed him is the broadening of the application of the processes to kinds of change other than speciation, because that is exactly what he himself was doing at that time. He was working on another book, entitled *The Descent of Man (and Selection in Relation to Sex)* (published in 1871), in which he broke his self-denying stance not to apply evolutionary thinking to humans, kept to carefully in *The Origin*. As the title suggests, this was an extended essay on humans, the human mind and evolution. In the section on language, Darwin drew an extended parallel between the formation of different languages and the formation of different species, applying both the terminology of his evolutionary theory and the processes inherent in it. He quoted approvingly the idea that 'a struggle for life is constantly going on amongst the words and grammatical forms in each language', and concluded that 'The survival or preservation of certain favoured words in the struggle for existence is natural selection.' That is as outright an application as you can get of a selectionist evolutionary theory – Darwin's own theory – to a realm separate from speciation. (Around a century later, a school of linguistics, as well as evolutionary modellers such as Luigi Luca Cavalli-Sforza, resumed the analysis of language change within this framework.)

A few years later, in the 1880s, the processes of evolution were shifted into the realm of the mind by Huxley and the great American philosopher-psychologist William James. Huxley reasoned that 'the struggle for existence holds as much in the intellectual as in the physical world', while James presented a prolonged analysis of individual creativity in general in terms of the Darwinian processes of evolution. It was not long before the physicist Ernst Mach and the mathematician Jules Henri Poincaré were using similar analyses to account specifically for scientific and mathematical creativity. Then, as the 19th turned into the 20th century, the psychologist James Mark Baldwin extended the application of evolutionary processes to both associative learning and individual development more broadly (which he called organic selection), but with an emphasis on psychological development. The latter was then expanded upon from the 1930s to the 1970s, though with a Lamarckian twist, by Jean Piaget.

Although science is a specific culture, Poincaré and Mach wrote about individual creativity in terms of the role it plays in science. Probably the first application of the processes of evolution to the way in which science works as science, and hence its first application to a specific culture of any kind, was by the philosopher of science Karl Popper in the 1930s – certainly Popper claimed to have been first in this regard, and I know of no one who came before him. In a sense, then, Kroeber's strange pronouncement that it was 'somebody's business to deal scientifically' with culture had already been anticipated by Popper, albeit within the confines of the particular culture of science. The first application of evolutionary processes to the understanding of any and every culture that this writer knows of was made in 1956 by the anthropologist George Peter Murdock. Shortly after this, the psychologist Donald Campbell began a nearly four-decade-long series of publications pursuing the analysis of individual learning and thought, as well as culture, within the framework of the universal processes of evolution. The phrase 'universal Darwinism' itself was coined in the early 1980s by Richard Dawkins. The work of Campbell inspired others to think along similar lines. A major review by Campbell in 1974, centred on Popper's work, contained just over 120 references covering the previous 100 years. By 1987 a further 500 papers on the topic had been published – something like a 40-fold increase in

publications per year. Clearly, the application of evolutionary processes to phenomena other than those addressed by Darwin in 1859
had become something of a growth industry.

Let's be clear what those processes are. In an influential paper
published in 1970, the American evolutionist Richard Lewontin considered Darwinian theory to be broadly made up of three principles,
which, taken together, drive the whole of evolution. (There is a matter
of terminology to bear in mind here: Lewontin uses the word *principles*
to mean subprocesses, the component elements of a process.) The first
principle, or subprocess, is that of phenotypic variation – differences
in phenotypic structure and function. The second is differential fitness,
corresponding to filtering by selection, and hence the differential survival and reproduction of variant phenotypes. The third is the principle
of heritability, whereby the traits contributing to the fitness of parents
are likely to be inherited by offspring whose own fitness will thereby
be enhanced. Variation, selection, and the transmission of selected
variants (heritability): run these principles, these subprocesses, as a
dumb and endlessly repeating algorithm, and any population will be
transformed by evolution.

What Lewontin then pointed to was the generality of these principles. They need not be confined to the Darwinian process of biological evolution. The mechanism of inheritance is unspecified. It could be
genetic, and in the case of biological evolution, of course, it is. But
it could also be cytoplasmic, involving cellular material outside the
chromosomes, or it could be cultural. The only thing that is specified
is the correlation between what the 'parent' has and some feature of
the 'offspring'. Variation could be driven by different mechanisms,
and the reasons for differential fitness, i.e. the nature of the selection
filters, are dependent on the nature of the system. What the unit is that
shows variation, and what constitutes the population, likewise depend
on the system in question. In the Darwinian case of biological evolution
leading to adaptation and speciation, the mechanisms are reasonably
well understood. What Lewontin was doing was emphasizing the
possible generality of the overall process without tying it down to
specific mechanisms. He was saying, in modern terminology, what
Huxley, James and others had been saying for over a hundred years.

Consider also the way the immune system works, which acts to protect us against bodily invasion by foreign organisms and other pathogens. The immune system is often portrayed, with good reason, as being at least as complex as the nervous system, running in parallel with the latter to protect the body from the vagaries of a dangerous world, precise knowledge of which cannot be supplied by the slowly moving main evolutionary programme (see chapter 2). That the immune system might be operating in a way best modelled in broad functional terms by the subprocesses that comprise biological evolution gave a considerable boost to advocates of universal Darwinism.

In the 1950s, riding on the wave of discovery generated by the Watson and Crick description of the structure of genes and the subsequent explosion of knowledge about molecular biology, Niels Jerne in Europe and MacFarlane Burnet in Australia began to model the way a part of the immune system, the humoral immune system, works as a kind of Darwin machine. The humoral immune system functions to bestow immunity against pathogens. In broad outline, it does this by generating a massive diversity of cells called lymphocytes. When one of these cells, with a specific antibody receptor structure, which varies from lymphocyte to lymphocyte, matches and binds with a particular molecular configuration of an antigen – which is a foreign cell or macromolecule that is not part of self – called an antigenic determinant, this triggers clonal selection of that antibody to help combat and destroy the invading antigen. After the antigen has been destroyed, large numbers of antibodies remain as 'memory' cells, thus conferring immunity, for varying periods, against subsequent invasions by the same antigen (i.e. pathogen). Such variation, selection and heritability transform our immune systems in response to the various antigens, such as bacteria and viruses, that we encounter during our lives.

The selection model of immune-system function was quite different from previous ideas about how the immune system might work, which were essentially instructionalist and Lamarckian. Lymphocytes had been thought of as being somehow malleable and moulded by antigens after initial contact with them to form structures that conferred immunity. Following Jerne and Burnet, immunologists realized that immunity arose from the selection of lymphocytes with variant cell-membrane structures that existed prior to the encounter with a pathogen. The immune system is a Darwin machine – and this is not hand-waving speculation. The entire and considerable weight of immunology underpins this understanding. Furthermore, the immune system, as an instantiation of the processes of evolution by selection, is a system operating within organisms, and employs mechanisms which are different from those of biological evolution. Discovering these facts gave great impetus to universal Darwinism.

It was only a matter of time before the idea was applied in earnest to culture and cultural change. Murdock may have been the first, but his approach was very broadly descriptive and seems to have had little impact. It is the quite separate writings of Campbell and Dawkins that have led to more detailed analytical models of culture and how it changes. In each case, a form of universal Darwinism lies at the heart of what is being proposed.

Modelling co-evolution

In 1973 Cavalli-Sforza and Marcus Feldman of Stanford University introduced the mathematical modelling that has come to be known as gene–culture co-evolutionary theory. Initially abstract, the modelling presented a formal method of linking cultural change on the one hand and the interaction of such change with genetic variation on the other. The classic case to which Cavalli-Sforza and Feldman subsequently turned their modelling was lactose tolerance, because this is a particularly clear-cut instance of biological evolution and cultural change interacting with one another. After being weaned, the majority of humans have difficulty absorbing lactose, which is a sugar contained in the milk of mammals. This is because the enzyme in the gut that allows babies to deal with lactose becomes deactivated after weaning, and the illness that results thereafter from ingesting milk can be serious. Such inability to cope with lactose also deprives people of a potentially nutritious source of food from their own livestock or that of their neighbours or others with whom they might trade. However, in some northern European populations a genetic mutation that prevents the deactivation of the relevant enzyme has been driven to fixation. Worldwide, the number of individuals affected is small, yet they enjoy obvious nutritional benefits – which is why the mutant gene was selected for. (People unable to digest milk are still able to consume milk products that have been processed to reduce their lactose content. Hence the popularity of foods such as yoghurt and kefir in parts of the world, such as the Middle and Far East and Africa, where most people are lactose intolerant.)

The anthropologist William Durham notes that an additional advantage of lactose tolerance is that lactose enhances calcium absorption, which is crucial for many bodily functions, not least the transmission of nerve impulses between nerve cells. Calcium deficiency can play havoc with brain functioning. Vitamin D, which is partly manufactured by the effects of sunlight on skin, is another important facilitator of calcium absorption, so for people living in places with relatively little sunlight, such as northern Europe, being lactose tolerant is doubly advantageous: milk can be consumed for its generally nutritious quality, and calcium absorption, which could suffer because of a lack of vitamin D, is given a boost.

The reason for Cavalli-Sforza's and Feldman's interest in this problem lies in its epitomization of the way in which cultural practices of animal domestication and husbandry, as well as methods of processing and preparing foods, have interacted with genetic changes in restricted populations of people. This kind of interaction is the key to such modelling, which makes two assumptions. The first is that genetic variation influences the likelihood that a particular trait, which could be a physiological characteristic, such as being able to absorb lactose, or a behavioural or psychological disposition, is transmitted between individuals. This does not mean that the genes in question are the main factor affecting that trait – whatever 'main' might mean in this context. And, of course, no trait can ever be caused by genetic variation alone because genes do not function within a vacuum. Genes are, though, as noted in previous chapters, the irreducible starting point of development. The second assumption is that the probability of an individual exhibiting a certain behaviour, holding a belief or value, or possessing particular knowledge, depends on other people, such as parents, peers or teachers, exhibiting that behaviour, holding that belief or value, or possessing that knowledge. In other words, these are cultural traits. Hence the phrase 'gene–culture co-evolutionary' theory.

Many human phenomena have been modelled in this way. Apart from lactose tolerance – which is the result of genetic variation on the one hand, and dairying, food-processing and cooking on the other, all of which are consequences of interaction between people within a culture – subjects as varied as the transgenerational maintenance of sign language (the cultural component) in communities affected by hereditary deafness (the genetic component), and mating preferences (the genetic component) in respect of personality traits (the cultural component), have been modelled. By making assumptions about the rates at which genetic and cultural change diffuse within populations of specific sizes, modellers have been able to get a quantitative handle on how human biology and culture interact.

Just five years after the first paper by the Stanford University scientists, another intellectual partnership, this one between the Americans Robert Boyd and Peter Richerson, began to develop conceptually similar models of what they called dual inheritance. The imagery is slightly different, but the issues addressed were identical. Boyd and Richerson placed a slightly greater emphasis on individual behaviour caused by two systems of inheritance, one biological and one cultural, and on how culture can result in group selection (more of which in chapter 6).

The really important feature of all such modelling is that it can capture the dynamics of the changes in both genetic and cultural structures of populations, and so can make predictions, when important questions are beyond the immediate reach of empirical examination. For example, in a study listed at the end of this chapter, Kevin Laland, Jochen Kumm and Marcus Feldman, making a modest number of assumptions about the genetics of sex determination and the cultural practices that do or do not discriminate against the sex of children, were able to make predictions about how sex ratios would change over time. One cannot experiment on humans when it comes to the matter of their breeding patterns or how they treat their children, nor can one simply rely on demographic data for answers, because collecting such data would take a long time and these are pressing problems. There is good evidence that as yet unknown genetic factors can significantly distort the primary sex ratio (that is, the sex ratio at conception) away from the ratio of equality that would be expected if the only thing determining the sex ratio was the way the X and Y chromosomes segregate during the formation of sperm. Other distortions in sex ratio may result from events between conception and birth and after birth. While purely biological or medical factors can be responsible for such distortions, cultural practices are just as likely to be the cause, and often much more so. Infanticide, report Laland et al., is practised to a greater or lesser extent in many societies. In some 36 per cent of pre-industrial societies it is common practice. The killing of infants, however, need not be direct. Neglect, gross underinvestment and abandonment have the same effect. And when and where such practices result in high mortality rates, it is girls who are overwhelmingly discriminated against. A study in India in the 1970s showed a consistent male bias in population trends over the previous century. In some regions of the world the distortion in child deaths 'can be as extreme as 5:1, which represents the killing, abandonment, or premature death of nearly 80 per cent of female children'. Advances in medical technology that allow accurate prediction of the sex of a foetus have led to sex-selective abortion, which increases the potential for bias. And whilst such sex-biased behaviour is not as common in Western industrialized society as in some other parts of the world, there is some indication of parental desire in both Europe and North America to control the sex of children as medical technology advances apace.

Well, these are clearly matters of great significance, and the ability to model and predict the effects of such biasing in sex ratios, both genetically and culturally caused, on future populations is valuable, to say the least. One of the most interesting features of such modelling is that the ways in which genes and cultural practices interact are complex and not intuitively obvious in outcome. Interestingly, it is not invariably the case that where girls are discriminated against, the sex ratio is always going to be distorted in favour of males. Much depends, for example, on whether the society is patrilineal or not, and whether it is the father or mother who is the main proponent of biasing the sex ratio of the family. Another complicating factor is whether or not parents adjust the size of their family following the premature death of a child or children. The underlying genetic causes of biased sex ratios can interact vigorously with cultural biasing to produce rapid changes in population structures. In such models the devil is in the detail. Small changes to values in the equations can have disproportion-ately large effects. The more these models are developed, the greater our knowledge of exactly what information needs to be gathered empirically so the models come closer to reflecting events in the real world. Sure, models simplify. But that is the point of models. They are developed precisely because gathering in the totality of events in the world as they occur in real time is too costly, too complicated and too slow to give us an understanding of what is happening. As Laland and his colleagues note, 'substituting a complex model for a complex world does not aid our understanding. When a simple model accurately predicts aspects of a complex world, we have gained some insight into what the important parameters may be.'

One of the advantages of such analytical methods is that detailed knowledge of the mechanisms underlying either genetic or cultural change is not necessary. Both Cavalli-Sforza and Feldman and Boyd and Richerson, as well as other modellers, do actually pay some attention to such mechanisms – especially those mediating cultural change, since these are the ones that we know so little about. However, it has long been a matter of contention as to just what detail of knowledge is needed before modelling can proceed. If one looks to population genetics, which was successfully modelling changes in the genetic composition of populations decades before the details of just what genes are and how they work became known, the answer is that very little need be known of mechanism. For example, Boyd and Richerson point out that when a child copies the behaviour of others, what is really important for the model is whether he or she copies the behaviour of a single exemplar or averages across the behaviour of multiple exemplars. For population-level modelling, this is a crucial issue. But how the copying is actually done, and the precise details of how the child chooses whom to copy, is another matter. True, when we understand these details of mechanism we will be able to give modellers more accurate information from which to calculate the probability of a trait being transmitted between individuals, or the degree of change that might occur during such transmission. But as long as modellers can hazard reasonably accurate guesses about such probabilities, the models will work reasonably well. Which, to repeat, is all one wants from a model.

One of the criticisms frequently levelled at biological approaches to culture is that being excessively gene-oriented, biologists too easily think of culture as being so constrained by genes as to be, in that infamous phrase, 'on a leash'. Well, until such time as culture, by way of biomedical science, has reached the point where people no longer die, it will, in an obvious sense, be on a biological leash. But this is no ordinary leash, because if, through cultural developments such as the invention of the chain saw and the creation of economic inequalities that force poor people to destroy vast swathes of the world's rain forests – if, by these cultural means, we are gradually depleting the planet's resources to the point where human biology cannot cope with conditions on Earth, biology is also on a leash and what is doing the leading is culture. The leash, as a causal force, is being tugged in both directions – it is a two-way street of causes. The whole point of gene–culture modelling is it recognizes that social forces and genetic variation needn't be directly tied to each other, but are reciprocally linked through the dynamics of their interaction.

Perhaps the best image to take from gene–culture co-evolutionary modelling is of two parallel tracks of evolving processes, the biological and the cultural. The forms of evolution of the tracks may or may not be identical in terms of process. What is certain is that quite different mechanisms underlie each track. But the tracks are not independent of one another: they interact and influence one another. They do this through lines of force, through causal connections, that link them and mediate the interactions between them, and these lines of force are the mechanisms of human intelligence.

The 'new' science of memetics

In the summer of 1999 about 50 biologists, social scientists and philosophers gathered at King's College, Cambridge, to consider the question 'Do memes account for culture?' – memes being putative cultural analogues of genes, and memetics the science of such entities. Among those attending were some of the most fervent supporters, indeed *the* most fervent supporters, of memetics – including the person who started it all, Richard Dawkins of Oxford University. The views of the others present ranged from acceptance that memetics had a significant future, if not in accounting for all of culture, at least in being an important theoretical advance which, when cashed out into empirical science, would prove a valuable component in a science of culture, to condemnation of memetics as a set of trivial notions irrelevant to any serious approach to social science. It is doubtful anyone left the conference with views different from those with which they had arrived. Memetics provides a good example of how people of approximately equal intelligence and knowledge can hold entirely opposing views. This is not the first time this has happened, nor will it be the last. But when it does occur it is one of the mysteries, and delights, of science. It arises from small differences between people in their knowledge, beliefs and aims that rapidly gear up into seemingly unbridgeable differences in scientific stance. One possible advantage to a future science of memetics might be better understanding of how this occurs, because, as with almost every other position taken up in this book, I make the assumption that the key to the future success or failure of memetics lies with psychology, specifically an understanding of certain cognitive mechanisms. The ways in which these cognitive mechanisms mesh with one another drive people to very different places in conceptual space. We need to know how this happens.

Let's begin at the beginning – actually, at two beginnings. The first is the appearance on Earth of a form of information transmission of massive significance. The second is a particular formulation of universal Darwinism. Turning to the first, we have already considered how social learning, as a means of acquiring information from others who obtained it from yet others before them, is practised both by chimpanzees and by various species of songbird. This slightly spoils the clean line of the story of the evolution of culture. As we know, culture is many things, and chimpanzees and songbirds do have culture of a kind. But, as is obvious, it is culture of a kind that has not changed the world. In the end, while birds sing beautifully and chimpanzees have some interesting cognitive features, they have not come to dominate the planet in the way humans have. Indeed, there is a real danger that wild chimpanzees will be driven to extinction by human activity in the next few decades. Whatever is different between human and chimpanzee, culture plays a huge part in it.

Sometime in the last few hundred thousand years, the human capacity for information transmission that by-passes genes crossed some threshold and the human capacity for culture became a power beyond that of any other creature in the history of life on Earth. The biologists Eórs Szathmary and John Maynard Smith were slightly off target in focusing only on the evolution of language as the crucial event here. But that, in a sense, is trivial. What matters is that they are absolutely right in pointing to the enormity of the importance of the evolution of the human capacity to transmit information extragenetically. This was an event as important as any other major evolutionary change in the three-and-a-half billion years of life on our planet. This is not simply because it involved our own species and led eventually to our being able to change our planet in a way that no species has ever been able to do before – and probably for the worse. Human extragenetic information transmission is a major systemic change in how life works. Something changed in our brains and our cognition, and that change was the crucial trick that gave rise to human culture. The change involved our ability to transmit information between ourselves independently of genetic lines of communication, an ability distinctly different even from that of the few other species of animal that can in some small measure do what we do.

The second beginning was the formulation of a universal Darwinism that operates within human culture – that *is* human culture. As already mentioned, Murdock was the forerunner in this, as far as I am aware – there may have been others. But it was Dawkins who really put it on the intellectual map. In order to understand how he did this, we need to make another small diversion into a particular formulation of universal Darwinism – that is, into the realms of replicator and vehicle theory.

William Hamilton's insights into inclusive fitness, and G. C. Williams' dictum that 'the real goal of development is the same as that of all other adaptations, the continuance of the dependent germ plasm', really did change evolutionary theory in a big way for those who followed them. People have always asked 'What is life all about?', and answers, ranging from the grand designs of God, through some kind of primal drive to survive, to evolution, have generally invoked some kind of inexorable force leading towards the creation of human beings.

Others, of course, have answered that life isn't about anything. Now we had important biologists giving us a different answer. Life, they told us, is about genes and their maintenance and continuity. This is a queer notion, a kind of metaphysical genetic reductionism, but it is an odd answer in part because it is an odd question. In any event, it is an answer that others enthusiastically embraced. E. O. Wilson insisted that 'the organism is only DNA's way of making more DNA' (DNA, remember, is the complex molecule of genes), and for Dawkins 'all adaptations are for the preservation of DNA; DNA itself just is'.

A slightly less odd, if more abstract, take on all this is by way of replicator theory. Some things are able to make copies of themselves, i.e. they can replicate, and when conditions are right this is what they do. There is nothing mysterious about it. The ability to replicate is inherent in the structure of such entities. It is, in fact, a rare structure and a rare ability. In sexually reproducing creatures, such as ourselves, offspring can never be replicas of parents. Gametes, or sex cells – the products of a type of cell division known as meiosis – are never replicas of the parent cells; and 'ordinary' cells – products of a kind of cell division known as mitosis – are not actually replicas of one another either. (Replicator theory takes, or should take, the word *replication* seriously. The copy must be a very good one. Mitosis does not produce identical daughter cells – similar, yes, but not identical.) But what *are* replicas in both kinds of cell division are the genes the cells contain. The very structure of DNA allows the molecule to unzip, one helix separating from the other, and then reconstitute itself perfectly as a double helix almost every time cells divide – almost, because tiny errors do inevitably occur, and these accumulate as we age. So genes are the archetypal biological replicators. They replicate because that is what they do when conditions are right, and those conditions are in effect the resources necessary for replication.

Now, if there are different replicators that overlap in terms of the conditions that are right for replication, they will 'compete' for those resources. There is nothing mysterious about this either. There is no intentional agency involved. The replicators are not selfish or aggressive. They are merely different replicating structures interacting with the surrounding conditions, including the existence of other replicators. And these conditions become selection filters, because some replicators copy themselves more effectively under certain conditions than do others. Some replicators are directly exposed to these conditions of selection. Others, as a result of the competition for the conditions best suited to their replication, evolve a kind of protective cloak, the phenotype, which is a non-replicating expression of their chemistry but which advantages them in the competition for the conditions necessary for replication. So, the genes are replicators, and the phenotypes, you and I, are what Dawkins called the replicators' vehicles. The word vehicle, though, is redolent of something under the control of something else, and Dawkins used this kind of imagery to suggest that it's the replicators that are really in charge of everything we do. But as we saw in chapter 2, any creature that has an evolved intelligence has, in effect, an additional layer of causation for what it does and how well it survives and reproduces. Indeed, some argue that even in the absence of intelligence, such are the complexities of development and of the intracellular and intercellular control of organisms, that genes cannot really be seen as being in charge of any organism.

In either case, vehicle is the wrong word. Over 40 years ago, before the work of Hamilton and Williams, the British developmentalist and evolutionist C. H. Waddington pondered the relationship of genes, phenotypes and evolution and concluded that genes are uninteresting chemical memory stores and that it is the phenotype, which he called the operator and which interacts dynamically with the world, that is what is really interesting. He vested the operator with causal power, but a more dynamic causal power than the 'boring' genes have. Decades later, the philosopher David Hull, pondering in turn the work of Dawkins, came to a view similar to that of Waddington. For him the word vehicle conjured up too much an image of passivity and being controlled rather than having some share in the controlling. And since the dynamic interactions that promote replication are, in fact, the subprocesses of selection, he substituted *interactor* for *vehicle*. The interactor is what selection acts upon, and those interactors that pass through the selection filters do so to the benefit of the replicators that designed them.

Again, do not read intentional agency into the word *designed*. It is used simply to refer to a massively concatenated sequence of selection events sifting and sorting through the variations thrown up by communities of replicators interacting with each other in such a way that variable phenotypes result. It no more implies agency than does Dawkins' 'selfish gene'. Dawkins used this metaphor to make a point, and as a kind of shorthand. The problem with metaphors, though, is that if they are vivid they can take over one's thinking. As Lewontin once said, the price of metaphor is eternal vigilance. In the end, replicators replicate simply because their structure is such that they do so under certain circumstances, and when other replicators are present under those circumstances interaction results in some replicators replicating more frequently or more accurately than others; and one way in which replicators replicate is by incorporating and constructing additional non-replicating structures – phenotypes or interactors – which support their replication. Replicators are no more actually selfish than is the moon made of green cheese.

In the last sentence of his 1976 book *The Selfish Gene*, Dawkins declares: 'We, alone on earth, can rebel against the tyranny of the selfish replicators.' Well, as we have seen in chapter 2, any intelligent animal is not wholly in thrall to its genes, so we are not alone on Earth in this regard. But Dawkins was talking about something else. It was not intelligence in general to which he was referring, but a specific aspect of it, namely culture. And in his last chapter, he raises the interesting question of whether human culture could be seen as being constituted of another form of replicator – cultural replicators, cultural analogues of genes. He proposes the name 'meme', because, among other things, it resonates nicely with 'gene'.

If cultural replicators, or memes, do exist, they behave like biological replicators, or genes. Under certain circumstances they replicate, because that is what they do, and may interact under those circumstances with other memes such that some replicate better than others – in essence, they compete for successful replication. So, in addition to genes being conserved by replication and transmission through the reproduction of organisms, memes, too, are conserved by replication and transmission: 'Just as genes propagate themselves in the gene pool by leaping from body to body via sperms or eggs, so memes propagate themselves in the meme pool by leaping from brain to brain via a process which, in the broad sense, can be called imitation.' The examples of memes Dawkins gave are 'tunes, ideas, catch-phrases, clothes fashions, ways of making pots or of building arches'.

Dawkins' formulation of cultural evolution cast in the form of universal Darwinism based upon replicator theory is a rather more detailed case than that made in the 1950s by Murdock; and it was tacitly taken up by gene–culture co-evolutionary theorists without too much fuss. But really, through the 1980s and early 1990s, the whole notion of memes as cultural replicators lay fairly low on the scientific scene, with occasional eruptions into visibility with the writings of the likes of David Hull. Then suddenly, from about the mid-1990s, memes started to attract wider attention. The new science of memetics was born – well, maybe – and conferences like that at King's College, Cambridge, were held. Whether memetics will blossom as a science of culture is debatable, as we shall see, but blossom as something it certainly has. Even the influential popular science magazine *Scientific American* recently carried an article declaring the importance of memetics along with some accompanying commentaries questioning its basic assumptions. So let us see what memetics is really about and what is wrong with it, at least in its present formulation.

Those who subscribe to the new science of memetics are overwhelmingly in favour of memes being understood as replicators in the same way that genes are replicators. They are also all for its substrate-neutral formulation. Using Dennett's distinction, memetics is about semantics (meaning), not syntactics (structure or mechanism). Perhaps because cultural evolution is seen primarily in terms of meanings, or meanings embodied in artefacts, there is also general agreement that it takes place orders of magnitude faster than biological evolution. Most memeticists are also none too bothered about what connections, if any, there are between cultural and biological evolution – which makes them quite different from gene–culture co-evolution modellers. And they seem to support imitation as the general means by which memes pass from mind to mind. Replication, substrate neutrality, the irrelevance of causal connections with biological evolution, rapid evolution, and reliance on imitation are what characterize memetics at the start of the 21st century. We shall examine each of these in turn. Before we do so, however, it will be useful to remind ourselves that, *pace* Lewontin, cultural change can be seen as cultural evolution within the more conventional formulation of universal Darwinism. Something about culture – loosely, cultural units – shows variation; some of these variable cultural forms are selected for over others and propagated and conserved through time, whereas others are less strongly selected for, or even selected against, and eventually become culturally extinct. Variation, selection and transmission: the net result is descent with modification within the cultural realm.

OK. But what *are* memes, really? For Dennett they are 'the units [that] are the smallest elements that replicate themselves with reliability and fecundity'. That second condition, fecundity, seems perhaps unfairly to burden the notion. If something copies itself with extreme reliability but does so only occasionally, why should it not be a meme of low frequency? Moreover, in a book that had a profound influence on a generation of biologists, G. C. Williams, in 1966, defined the gene abstractly as an 'entity that must have a high degree of permanence and a low rate of endogenous change, relative to the degree of bias'. At the King's College conference, David Hull, following J. S. Wilkins, offered: 'A meme is the least unit of sociocultural information relative to a selection process that has favourable or unfavourable selection bias that exceeds its endogenous tendency to change.' His point was that if we criticize this definition of a meme because it is absurdly abstract, we are being inconsistent in our high expectations of memetics because no one howled with derision at Williams. Quite the contrary. What Hull was saying was stop being so critical and nit-picking and let's get on with the doing of the science to see if it works. After all, if nothing else, memetics offers us a potential unit of culture which, in principle, is measurable in some way – and science lies in the measurable.

But as soon as we start to look for examples of what a meme is, we run into the problem of substrate-neutral formulations. If, as Dawkins suggests, a tune is a meme, is it the tune as sung and heard, or is it the brain state of the singer as the song is sung? In other words, which of these is the replicator and which the interactor? (Modern technology can be ignored here, for it really does not have anything to do with a science of memetics, which must account for all of human culture, including prehistoric culture (before 5000 BC), so forget about recent innovations such as recorded script, tapes, CDs and computer hard disks.)

This is a very difficult question to answer. A tune is the result of a sequence of neural network states being played out somewhere in the brain and translated via a flow of nerve impulses to the vocal cords, which vibrate in another sequence of events to produce the sound of the sung tune. The sung tune is a transient phenomenon that is heard by someone else by being coded into another sequence of nerve impulses, which are conducted along the auditory nerves into the auditory regions of the hearer's brain, decoded, and then stored as a set of the hearer's neural network states. Insofar as one can get away with not specifying exactly what these streams of nerve impulses and neural network states are, or even where they occur, one can maintain substrate neutrality. There is, however, a very strong implication of neural mechanisms. Are these implicit mechanisms the replicator, or is it the sound waves created by the singer and heard by the receiver? Since it is a replicator we are looking for, the answer seems to be the unstated mechanisms. The sung song would appear to be the interactor, because this is what the hearer selects, and selection, generally, is held to act on the interactor. Replicators, remember, just make copies of themselves.

But if that is the answer, we immediately run into another problem. I do not walk about all day with the memory of a specific tune prominent or even imminent in my mind. If I want to remember a particular tune, say the national anthem of Britain, why then, I literally re-member it. I take the potential of the tune and put it together by a process of re-constitution or re-call. All memory works this way, even if we know little of the mechanisms involved. All memories are potential brain/mind states that become actual during re-collection or re-membering. This does not only apply to memory for tunes. Almost no one carries round the conscious memory of the fact that Abraham Lincoln was assassinated in a theatre. But if I ask you how Lincoln died, you re-member that information. And because the brain is a dynamic organ system, the activity of which is constantly changing, and which is never in a state of inactivity (except after death), the tune as I re-call it now, or the memory of how Lincoln died, is re-membered upon a slightly different background of overall neural network activity than will be the case should I remember it an hour, a day or a week later. Worse still, singer and hearer have different brains, each with a different history and each with different neural networks in different states. The networks may be in the same part of each brain, but their states cannot be the same. That means the song is almost certainly stored in the memory of the hearer in a different, even if only slightly different, set of neural network states from that of the singer. From this, one can only conclude that if brain states are the replicators, then we are not dealing with replication as we know it from genetics. It is very messy replication.

This is precisely why people like Dennett say we have to keep memetics substrate neutral. We can all learn simple tunes by hearing them and can then reproduce them with accuracy. It is the tune that is replicated, not the brain states. But if the tune is the replicator, then what is the interactor? What is selection acting on? It could be argued, and has been, that in looking for separate replicators and interactors we are being held too strongly to replicator theory as applied in biology, where it is usually easy to distinguish between genes, genotypes and phenotypes. There are known instances in biology of replicator and interactor being one and the same thing. Examples include certain intracellular structures, such as ribosomes (small particles of RNA and protein in which protein synthesis takes place) and prions (specks of protein presumed to be the cause of degenerative brain diseases like BSE, so-called mad-cow disease, and its human form, Creutzfeldt-Jacob disease). So could it be that in the realm of culture, the distinction between replicators and interactors is unnecessary, even distracting? Well, perhaps. But if so, those few biological exceptions notwithstanding, this is a huge disanalogy with replicator theory in biology. Is it the same theory?

The sources of variation are less of a problem. If brain states of some kind are the replicators, the dynamic nature of brains and the differences between brains will supply variation aplenty. Some replicators will be less variable than others. *God Save the Queen* comes out pretty well the same whoever is singing it. But other memes are highly variable. My meme for Shakespeare's *King Lear* will only approximate to yours, and both will be rather different from what appears on the printed page. In a famous series of experiments carried out by the English psychologist F. C. Bartlett in the 1930s, a story was told to one person, who, after an interval, recalled it for someone else, who then retold it to a third person, and so on. This is the children's game of Chinese whispers. Bartlett carefully tracked the changes and was able to document specific ways in which the story was degraded from its original version to a rather conventionalized and dull husk of the original. In real life, degradation is to some extent countered by people spicing up stories and gossip with invented or imagined details, but, of course, this results in further departure from the original. This is a commonplace of ordinary social interaction, and it raises the problem of excessive variability. Dawkins had argued that in biology good replicators have fidelity, longevity and fecundity; that is, they are able to copy themselves with accuracy, maintain their structure for long periods of time and reproduce themselves in quantity. Copying fidelity could, for many memes, be a problem. If narratives are memes, and they surely must be, fidelity will often be much reduced. Yet in some instances it will be remarkably good. If I tell my friends that the *Café Délicieux* on the Avenue de Gourmand serves great food and is inexpensive, by their actions of dining there and not at the restaurant just round the corner, copying fidelity is well maintained. It really seems to depend on the kind of meme one is talking about. Not all memes are equal in their characteristics. This is a very important point, to which we will return shortly.

At that King's College conference, the French anthropologist Dan Sperber suggested that many memes of the I-expect-you-to-contribute-to-the-costs-of-running-this-household kind gain a degree of reliable copying by the receiver using inference to determine the intended meaning of the speaker. This raises the spectre of mechanism again (specifically a theory of mind module, which will be considered in the next chapter). In other words, the brain employs general decoding processes, embodied in specific mechanisms, accurately to transmit and receive memes that have meaning. This must be correct. But it doesn't stop with the inferring of intentions. This is a meme that also carries a large load of prior knowledge, such as what a household is and that its running involves expense, which sender and receiver must both have. In other words, memes are psychologically very complex things indeed (something we will return to shortly). This does not count against replicator theory as a tool for studying culture, but it does mean that memetics is no easy or simple solution to the complexity of culture. It also means, as will be argued in the next chapter, that replicator theory in culture must make contact with the evolution of the human brain's capacity to gain knowledge of specific kinds and to make certain kinds of inferences. Memetics must take into account not just the proximate causes of how a cultural replicator gets from the mind of one person to that of another, but also the distal evolutionary causes of the mechanisms that allow this to happen. Memetics, if successfully pursued, will be a big and necessarily expansive area of science. This is not an argument against substrate neutrality. Processes really do matter. But, as with all of biology, it is impossible to keep mechanisms out of the story. At some point substrate neutrality must give way to embodied mechanism. And this has advantages, because knowing about mechanisms gives us a much clearer view of the range and dynamics of variation, just how selection works, and the fidelity of the transmission process.

What of the tendency of memeticists to seal cultural evolution away from biological evolution and see it as a self-contained system – as a system that is not held on a biological leash, to use that famous phrase? Using replicator theory in this way, but shifted into the realm of culture, seems an attractive proposition. If replicator dynamics, and only replicator dynamics, are what is needed to understand culture, then what we have is a general theory freed from biological constraints. But, as just noted, as soon as you bring mechanisms into the picture, you have to account for them as evolved structures of the mind. Furthermore, if the gene–culture co-evolution theorists are correct in their general architectural scheme of causal links, no matter how tenuous, between biological and cultural evolution, then using memetics as a device for countering the criticism from social scientists that biologists always insist on keeping culture on some kind of biological leash doesn't work. Much depends on what one means by a leash. As already pointed out, if culture can affect biology, as in the case of lactose tolerance in some human populations, it is not clear what is holding what on a leash.

Another common observation about memetics, though often made with regard to culture in general, is the rapidity with which cultural change occurs. This is so widely held as to be almost a cliché among both social scientists and memeticists. But is it true? In one sense, yes.

Clothes and other fashions change on an almost yearly basis. Where people shop and what they buy also seem to change fast. During the last five years, use and knowledge of the Internet seem to have spread phenomenally quickly. Well, in the latter case perhaps we are being influenced by the profoundness of the changes that are affecting our lives rather than their rapidity. Remember how in chapter 1 we noted a number of well-documented cases of biological evolution occurring in just a few generations over none too many years. That is about the same pace of change associated with information technology, at least as it impinges on ordinary people. But it is the word *generation* that matters here. All biological evolution is temporally constrained by generation turnover time. In humans that is somewhere between 12 and 14 years. So when cultural generation time is determined by social institutions like the great fashion houses of *haute couture* to be annual, then this microcosm of human culture is clearly moving fast.

However, much of human culture actually comes to us usually only once in our lifetime. We acquire the knowledge structures, values and beliefs that characterize a particular culture through a long period of childhood enculturation. The inheritance may be messy and spread out over many teachers and enculturators over a long time, but as a 'shot' of information transmission it happens just once – just like genes. In other words, some memes, important memes, have the same generation time as genes. That may go some way to accounting for the extraordinary persistence across time and generations of some features of world cultures. The great religions of the world are thousands of years old. Some may, just may, be waning in certain societies. Others are clearly waxing. It would be an unwise person who, in the light of recent history, predicted the rapid decline of the great religions.

There is another consideration here. If religion is a form of 'making sense' of a complicated and fearsome existence, and is present in varying form in most cultures, then there are features of cultures which are probably universal and certainly very ancient. Food preparation is another example. The surface details may change, but the core 'thing', preparing food, does not. It is older than recorded history. This is not the place to analyse the persistence of such features within and across cultures, but they make a general point, a point which also speaks to that fifth common assumption about memes. This point is that some memes are universal and others are not; some change fast and some change hardly at all; and often the memes that are changing fast lie on the surface of the memes that are changing slowly.

Let's now move on to the final assumption about memes, which is that their transmission is by imitation. The central tenet of universal Darwinism, remember, is that subprocesses of variation, selection and heritability will exist, whatever their medium of instantiation. In replicator theory, heritability is an essential part of replication, and replication implies some (unstated) degree of fidelity in the copying process. Now, when Dawkins, back in 1976, first raised the possibility of applying his replicator version of universal Darwinism to culture, he envisaged memes as leaping from brain to brain by way of imitation 'in the broad sense'. Well, Dawkins in 1976 was being creative beyond the wildest dreams of most biologists and no criticism should be made of his cautious choice of words. 'In the broad sense' means what it says, which is very little. It is utterly vague and can be taken to mean anything, which one can only assume is precisely what Dawkins intended over a quarter of a century ago. But with the recent revival of memetics, we find ourselves saddled again with imitation, and it still seems to be 'in the broad sense'. What was acceptable in the 1970s when new conceptual ground was being broken is not acceptable to social scientists in the 21st century.

The word *imitation* was introduced as having some kind of technical meaning into psychology over a hundred years ago by E. L. Thorndike, for whom it meant 'learning to do an act from seeing it done'. You see someone doing something, and then you do what they did as they did it. It sounds very simple, but as we will see in the next chapter, it isn't. Simplicity or otherwise apart, the meaning is quite clear. Input, along the visual nerves, of the action of another is transformed into a similar act by oneself. There has been much argument as to whether non-human animals are capable of imitation, and some of that argument has centred upon the point at which 'dissimilarity' of action rules out imitation. This is not the stuff of which great science is made, and it is quite clear that those creatures of protoculture, songbirds and chimpanzees, do somehow come to share similar behaviours. In the next chapter we shall see that whether it is imitation that is responsible is open to question, and that current evidence suggests it is not.

Nonetheless, these animals do get to share through learning what others have learned.

The continuity of stone tools through two million years of human evolution has two possible explanations: the repeated independent discovery by separate individuals of how to flake and shape rocks into cutting implements, or the repeated acquisition of such skill by observing others. The latter is a much more likely and reasonable explanation than the former. Whether that common ape ancestor of chimpanzees and humans of five-and-a-half million years ago had imitative skills anything like those of early humans is unknowable, if doubtful. Chimpanzees are nowhere near as skilled as we are in manipulating objects with their hands, and while evidence as to whether they can really imitate in the Thorndikean sense is conflicting, experiments indicate they cannot. There is no reason to think chimpanzees have lost a form of intelligence possessed by ancestral species, so that common ancestor was almost certainly not capable of imitation. That we are now capable of imitation, and that the continuity of stone tool usage over two million years is most easily explained as the result of a form of social learning, probably imitation, suggest there really should be no doubting that imitation, in the proper use of the word, has played an important part in human evolution. It may well have been absolutely central to the evolution of human culture, though we can probably never know this either. But what is clear, and eminently knowable, is that imitation is not central to human culture today. Yes, it has its small part to play. We do learn from elders and peers how to dress, tie shoe laces, wield forks and knives and hold a pen. But we do not acquire our native languages through imitation (see next chapter); we do not learn the basic knowledge structures of our cultures, such as what a shop is and how it differs from a school, by imitation; and we do not come to understand the basic values and beliefs characteristic of our cultures, such as what justice, God's law and the pursuit of happiness are, by imitation. The mechanisms – sorry to have to use that word again but we have arrived at yet another point where process alone has insufficient substance to carry the weight of argument and analysis – the mechanisms of motor imitation, language learning and knowledge gain are different in each case; as we shall see, they develop in the infant and child at different rates; and they are sited in different parts of the brain.

Why, then, are contemporary memeticists so wedded to the notion that imitation is the essential means of meme transmission? Well, one possible reason is that they just do not know enough psychology, and another is that they are not discerning in their use of language and are happy to employ the word *imitation* in so loose and sloppy a manner that it loses all meaning. But a more likely reason is that running the substrate-neutral position so hard simply blinds otherwise intelligent and knowledgeable people to differences when the processes are fleshed out with mechanism. There are, though, two other reasons for this lack of discernment, and both have to do with replicator theory.

The first of these is copying fidelity. There *seems* to be less room for copying errors and variation when one is dealing with simple motor acts than when one is dealing with complex knowledge structures or beliefs. This may be correct, though it is questionable. For one thing, fluctuating neural-network states are a universal condition. Whatever is being copied, be it a simple motor act, a sentence, a narrative or a belief, involves neural-network states fluctuating across time within individuals and differing between them. For another, anyone who has tried to teach others a relatively straightforward set of motor skills, like how to serve in tennis and where to place one's feet when playing a forehand or backhand so that the ball is correctly addressed, will know that people vary hugely in their abilities to acquire such skills – that is, to imitate in the proper sense. Some just never get it. Some get a bit of it but never reach a consistent level of skill. A few imitate well. Yet stop people in the street and ask them if they understand what a shop is and how it differs from a prison, and you will find that only disease or brain damage prevents every member of our culture from knowing these things. Variation in the latter case is far, far less than in the case of the transmission of motor acts. But even if there is some sense in which the possibility of error is greater with knowledge structures and belief systems than with motor skills, simply because the former are more complex and there is therefore more room for error, this does not rule them out as memes. Just because replicator theory has its origins in biological theory, where the archetypal replicators, genes, have astonishingly high copying fidelity, is no reason to suppose that similar fidelity is necessary when applying replicator theory in other realms. Indeed, it may be precisely such lack of fidelity in the copying process that is the source of culture's richness and complexity. Embracing less faithful copying mechanisms may make memetics a good deal more acceptable to social scientists than it is right now.

The second reason for such adherence to imitation is, perhaps, the idea already discussed – that cultural change is, supposedly, quicker than biological evolution. Imitation is supposed to be quick, clean and cheap – the mistakes of others will not be yours if you imitate what is successful in others – and can occur over and over, many times in a life, indeed many times in a day. This is, however, but a quarter truth. Some imitation, of the most simple acts, is quick. Some takes much longer. But more significantly, the most important things about culture are acquired across a period of years during the process of enculturation. We learn our native language, for example, and acquire a variety of knowledge and values and beliefs central to our culture just once, during childhood. The process of transmission for matters such as these is neither rapid nor repeated.

But this simply does not seem to be true, some will say. Fashions change rapidly, as do favourite recipes and favoured restaurants. Well, this is obviously correct. But when I tell a friend that the *Délicieux* is the best value in town, I am building a very small detail onto an already massive edifice of knowledge – what a restaurant is and what makes for good value – gained over the years of childhood and adolescence. It is only because of those years of slow enculturation that we can later transmit information so quickly and, yes, accurately between one another. The main message here is that there are memes and there are memes. Some, the deep-structure memes of any and every culture, are transmitted slowly, so slowly that the process usually occurs just once in a lifetime. These are memes of wide informational scope. They take in a great deal of knowledge. Others, the surface memes, are the social froth of daily life in any culture. They are memes of very narrow, specific informational scope: who is nice and who is not, where to get one's food and where to avoid, what clothes to wear and what never to be seen dead in.

Well, the idea of different 'levels' of meme, where *level* refers to the scope of information being transmitted, does have its parallel with replicators in biology. The homeotic genes, which are deeply conserved complex gene clusters concerned with the development of structures on a broad scale, similarly have wider scope than do structural genes which govern simple characters such as eye colour. All genes are not equal, so why should we expect all memes to be so?

Little has been said about the selection of memes, and in the memetics literature selection has not, so far, figured large. If the gene–culture co-evolutionary modellers' principal assumption about the interaction of biological and cultural realms is correct, then the very dynamic of this interaction is a source of selection. The domestication of animals formed the cultural environment within which the gene bestowing lactose tolerance was then driven to fixation; and the existence of dairying and the advantages of milk consumption then selected the processing of dairy products that reduce lactose levels in those populations intolerant of lactose. In essence, this is an adaptationist account that brings both culture and biology into the analysis. Another, more speculative, source of selection lies in human evolution for life in small social groups. This is an idea that will be pursued in the penultimate chapter of this book. In brief, what it raises is the possibility that we humans are predisposed by our evolutionary history to adopt certain memes rather than others. If this is so, here is another link between culture and biology. However, as the philosopher India Morrison points out, there are certain things that do seem to leap from brain to brain, perhaps in a manner analogous to the contagion of yawning and laughter, a contagion that seems to be without adaptive significance. These are catchy tunes, fragments of rhyme, gestures and postures, all of which, unlike laughter and yawning, are transmitted by acts of learning. What these have in common, suggests Morrison, is 'they tend to be brief, non-complex musical or spoken phrases with strong rhythmic and/or amplitudinal ratios . . . also they tend to possess a (sometimes overweening) motor component, creating the urge to move, tap, hum, sing, burst into verse . . .' If everyone is susceptible to 'invasion' by memes such as these, then it must be the case that we share brain structures causing this susceptibility. If that is so, part of the meme selection problem resides in inherited neurological characteristics.

Meme selection may also be importantly affected by forces that arise from within cultures, and by specific features of those cultures. For example, the similarity of newly arising memes to other already existing memes may significantly affect their selection. So too may how well such new memes fit in with existing high-frequency memes, or even how much they stand out from prevalent memes – a kind of novelty effect. Finally, since memes are supposed to be the units of culture, forces familiar to social scientists, such as status and group membership, may exert significant selection forces on memes.

In conclusion, just as memes are not all of a single kind, and just as their transmission between people is achieved by multiple means, so it is likely that there are different selection filters sorting and sifting between memes. This memetic pluralism may make the meme approach more acceptable to social scientists, if only because it allows in a little of the complexity that they know characterizes all cultures. In fact, insofar as both memetics and gene–culture co-evolution models are built on foundations of human intelligence, and given that intelligence, as outlined in chapter 2, is a semi-autonomous set of devices that frees human thought and action from the direct control of genes and allows in another layer of causation, social scientists are generally more welcoming of these biological approaches to culture than they were to the previous generation of ideas from sociobiology, which demanded a much tighter linkage between genes and culture than was acceptable. The latest ideas at least begin to allow for answers to the puzzles of war, celibate priesthoods, sex-biased infanticide and a host of other phenomena which defy explanation purely in terms of biological fitness. They also, as the anthropologist Maurice Bloch notes, expunge the implicit dualism that characterized much of classical anthropology – remember the culture-as-transcending-minds approach mentioned in the previous chapter.

Many social scientists, however, including those who, like Adam Kuper, welcome the presence of NeoDarwinian thinking in the social sciences, remain highly sceptical of either memetics or gene–culture co-evolutionary theory being able to deliver real insights into humans as creatures of culture. One reason is a 'So what's new?' shrug of the shoulders, a weariness with the ignorance of biologists of the history of the social sciences. Earlier generations of anthropologists were Darwinian, adaptationist and diffusionist in approach. The new theories coming from biologists are anything but new, and they raise no novel insights. All these social scientists see is old and not very good wine in new bottles. They object to the notion of memes because they cannot see what understanding comes from shoe-horning tunes and catch-phrases into the same category as religious precepts and rituals. These are, they assert, manifestly different things, and banging them together into the same box doesn't create understanding, it erodes it. They point to the presence of intentionality in the way that cultures change, a disanalogy with biological evolution that is so great as to make the exercise meaningless. They deplore the way in which biologists ignore the force of history in shaping cultures – though it must be said this is a queer criticism, because if Darwin taught us anything, it is, as Stephen Jay Gould has repeatedly pointed out, that history is important. They point to the differences between universal Darwinism as it works in biological evolution and as it works in cultural evolution. For example, there is good reason to believe that there is multi-parentage of memes, including peers as parents, and a possible blending of memes from these multiple parents. The gene–culture co-evolutionists explicitly recognize this, but the social scientists fret about such transmission differences introducing such enormous differences in the dynamics of inheritance and variation that it raises serious questions as to the value and validity of the supposed parallels of universal Darwinism operating in different realms. In essence, this is another version of the criticism that substrate neutrality is a dangerous way to proceed. It must, social scientists imply, be bolstered by mechanism. The semantics and syntax are irretrievably intertwined. In this they are correct, even if this is an interesting advance in social-science thinking.

But above all, social scientists deplore the atomization of culture by biologists. 'Unlike genes, cultural traits are not particulate,' writes Kuper. 'An idea about God cannot be separated from other ideas with which it is indissolubly linked in a particular region.' Judaeo-Christian monotheism is a different system of ideas from Graeco-Roman polytheism, which in turn is different from Hindu polytheism. So there is not some single, simple, discrete God meme. Anthropologists also deplore the 'deinstitutionalization' wreaked by biologists, who downplay or, worse, ignore the synergistic relationships between cultural knowledge and beliefs and the matrix of institutions they give rise to, which in turn strengthen, or in some cases change, those beliefs.

This is quite a catalogue of sins, though some may not be reasonable criticisms of what biologists are trying to do as they stalk this thing called human culture in an attempt to add to our understanding of it. I list them here not to refute them, rather as markers against which we might measure the chapters to come. Our next step is to turn to those mechanisms in which the processes outlined in this chapter must be embodied.

Suggested Readings

Aunger, R. (2000) *Darwinizing Culture: The Status of Memetics as a Science*. Oxford, Oxford University Press. (The edited proceedings of the 1999 King's College symposium, containing views for and against memetics.)

Ehrenreich, B. (1998) *Blood Rites: Origins and History of the Passions of War*. London, Virago. (A readable account of theories of war and an account of the human transition from prey to predator.)

Grossman, D. (1995) *On Killing: The Psychological Cost of Learning to Kill in War and Society*. New York, Little, Brown and Company. (Documents the reluctance of most people to kill others.)

Journal of Memetics. www.cpm.mmu.ac.uk/jom-emit/ (An electronic journal available free on the Internet for those who want to know the latest in memetic thinking.)

Laland, K. N., Kumm, J., and Feldman, M. W. (1995) 'Gene–Culture Coevolutionary Theory: A Test Case.' *Current Anthropology*, vol. 36, 131–58. (A specific example of gene–culture co-evolutionary modelling in action.)

Scientific American (2000), vol. 283, 52–61. (Contains an article in favour of, and commentaries expressing reservation about, the meme idea.)

5

Causal Mechanisms

Patience (also known as solitaire) is a card game in which one moves from an opening, semi-random, array of cards, the initial state, to an end state, which is the most orderly array of cards possible given the initial state, through manipulating the cards and the positions they occupy according to a limited set of rules. There was a time when this game was played only by hand with a real pack of cards, and the rules were contained in, and operated by, the mind/brain of the player. These days, the game is a standard feature of the games section of most desktop or laptop computers. This is yet another example of a process – which is what the game is as it moves from initial to end state – that can be embodied in more than one way. When I learned the game as a child, I did not, of course, know I was learning how to proceed through a process by way of a substrate-neutral algorithm. And if I had been asked to explain the game, it would have been in terms of that opening array of cards, the initial state, the rules by which the cards are manipulated, and the final state that is the goal of the game. For some scientists, that kind of description (by a child!) is what the biological sciences are about. Process is all. Well, as we saw in the previous chapter, process is certainly very important, and substrate neutrality is an especially interesting property of a process, signalling as it does the presence of general descriptive dynamics. But it is never enough because, within the realm of the biological, all processes are driven by specific causal mechanisms, entities which cause the processes to run. To repeat Dennett's words, mechanisms are the 'causal powers of the materials' used in the instantiation of a procedure or process.

So, while it is certainly significant that card games can be embodied in different causal mechanisms, that does not diminish the importance of being able to understand what those mechanisms are. After all, you can't throw a collection of threads, dead leaves and CDs of all last year's music bestsellers into a heap and expect them somehow to combine to produce a player of patience. Mechanisms are not arbitrary and they are not random. They must have very special characteristics in order to cause processes to occur, and what those characteristics are is an essential part of the causal explanation of anything biological, including the things that make humans special. It may well be that physics can dispense with mechanisms and be played out by abstractions like superstrings operating in an ethereal universe of multiple dimensions. The biological, however, including the human, exists within intermediate levels of size and force. The range from viruses to whales, from the order of DNA to the organization and connectivity of billions of nerve cells in the human brain, is the range of biological reality. Of course, this must find ultimate inclusion within the total range from the smallest subatomic particle to the largest entity, i.e. the universe (or many universes). This is the ultimate aim of the unity of science, and we can expect progress in this direction over the next century or so. But right now our goals within the larger project of an overall unity of science are much more humble and limited and have to do with marrying the social and biological sciences. Both involve levels of organization of great complexity, but always of this intermediate level of order embodied within touchable, edible, entities. So when someone says to me 'Give me a causal explanation of that child's behaviour – that's right, the one playing patience', an account cast solely in terms of processes is incomplete. We need to know about the entities inside the child's brain and mind that allow the processes to run and which are absent from our inanimate, arbitrary and random heap of threads, leaves and old CDs. Specifically, we need to understand what these mind/brain entities are, and how they are organized into causal mechanisms, because within the range of the biological, organization is all, to paraphrase William Calvin, the American theoretical neuroscientist.

A general framework for understanding psychological mechanism

Within psychology itself there is not much argument over the fundamental assumption that the mind is a set of processes run by mechanisms and that it is the job of psychology to work out just what those processes and mechanisms are. By extension, then, if culture is one of the manifestations of human intelligence, and if it is to be 'someone's business to deal scientifically' with culture, to return to Kroeber's strange statement, then understanding both the processes and mechanisms of intelligence is the way to deal comprehensively with a science of culture. In the previous chapter we concerned ourselves mainly with process. Here we turn to the crucial mechanisms – or what we think are the crucial mechanisms. Before we do, though, a few points must be cleared up.

The first is that readers new to psychology may wonder just how one looks for mind mechanisms. After all, it's one thing to envisage, say, the chemical mechanisms of liver function that clear toxins, such as alcohol, from the blood. This is as materialist and physical a process as one could imagine. With sharp enough eyes and small enough fingers, one could see and touch the cellular chemistry of the liver. But the mind? Isn't the mind something immaterial, something without substance – a quality which characterizes it as surely as material substance characterizes the liver? This is the mind–body problem touched on briefly in chapter 3 and dealt with just as briefly here. No one doubts the formidable difficulties faced by our need to relate the mind to the body, specifically the brain. There are a number of ways of doing this on offer and none is universally accepted. What *is* universally accepted is that the mind is the result of chemistry in the same way as is the liver. In some manner not yet properly understood, the mind and the brain are one thing. There simply is no other possibility. Part of the difficulty is that we have at present only the dimmest knowledge of how the mind is structured and how it works, and while we have much more understanding about the literal chemistry of the brain and its connectivity at a gross level, we have as yet no knowledge of the microstructural details of how sheets of hundreds of thousands of nerve cells with their tens of millions of interconnections are organized into structures that relate directly to mind processes and mechanisms. Nonetheless, the universal operating assumption is that one day in the future we will have this knowledge, and then we will truly be able to say that the mind is the same material thing as the brain.

'Well, fine,' you reply. 'I'm a tolerant person and accept your humble position of ignorance combined with faith in the material nature of all things and the ultimate triumph of science. But I still don't understand what form, exactly, a psychological mechanism takes, how certain one can be that a hypothesized mechanism does actually exist, and how such things differ from our intuitive, commonsense, understanding of how the mind works.' Let's deal with the last question first.

We are all of us very good at understanding the motives and perspectives of others. However we come to this knowledge (more of which later), we understand the advantage of friendliness over aggression, the importance of being able to anticipate the responses of others, and the value of reciprocity, among many other things that oil the wheels of social interaction. Given the daily needs of an intensely social species such as ours, this is unsurprising. While an appreciation of how we come to have such understanding is a legitimate and important aim of psychological science, the contents of such folk psychology are a trivial part of normal socialization. What is important is comprehending what underlies such understanding. (A brief note: What later in this chapter will be referred to as Theory of Mind is, occasionally, labelled folk psychology. This, of course, is horribly confusing to outsiders. When I use *folk psychology* here, it refers to the overall surface understanding each of us has of what makes other people 'tick'. It does not refer specifically to the technical mechanism by which we attribute intentional mental states to others.)

Take memory as an example. That we all have memory is understood by all normal people. It is a commonplace of folk psychology. That memory has a complex structure, however, is one of the important discoveries of 20th-century cognitive science. The possibility that memory might not be all of one kind was actually first raised by the American philosopher-psychologist William James in the 1880s. His distinction between primary and secondary memory, to use his terminology, was built on by the Canadian psychologist Donald Hebb over

60 years later. Hebb proposed that the first of these, primary memory, might be sustained by fragile and easily disrupted electrical patterns of activity between clusters of nerve cells, whereas the more robust secondary memory was the consequence of structural changes, such as growth, between cells. The well-known example of how we have to keep repeating and rehearsing a new telephone number just given to us by an operator as we key it into a telephone, how we fail if the number is too long and we have not written it down, and the way this contrasts sharply with the enormous quantity of information that we can remember without having constantly to repeat it, seems to support the distinction. Primary memory, now usually referred to as short-term memory or working memory, has a very limited capacity and duration. Secondary memory, better known now as long-term memory, has an almost infinitely large capacity. Experimental studies have pretty well consistently supported the existence of these two fundamentally different, though connected, forms of memory. For example, if someone is asked to listen to a string of unconnected words and then asked to remember them, there is a strong tendency to remember the last few words on the list better than those earlier in the list, which is known as the recency effect. But if a delay, of even just five to ten seconds, is introduced between hearing the list and recalling the items, the recency effect disappears, while there is no change in the likelihood that items nearer the start of the list will be recalled, and be recalled with the same frequency as when there is no interval between hearing and recall. An experimentally robust finding that is easily replicated, this is clear evidence supporting the existence of separate memory mechanisms.

This experimental teasing-apart of two different memory mechanisms is supported by studies of people who have sustained damage to their brains. Some patients have greatly reduced short-term memory – sometimes it seems to be entirely absent – with relatively intact long-term memory, while the pattern in people with damage to different parts of their brain indicates that they suffer from a reverse form of memory loss. Subsequent studies have indicated the likely presence of other differences within long-term memory itself. Memory for language may survive even the most global forms of amnesia, whereas the consequences of cerebral accidents (i.e. stroke) and other forms of damage show that loss of memory for words, referred to as category-specific anomic aphasia, can be remarkably specific (such as forgetting the names of fruits or vegetables while retaining memory for all other words), depending upon the site of damage. There are also differences between memory for meaning, called semantic memory, and memory for everyday occurrences, or episodic memory. Another commonly assumed difference is between implicit and explicit memories, which is sometimes mapped onto the difference between knowing-how, procedural (doing) memory (for skills like tying shoe laces), and knowing-that, propositional memory (for remembering Brussels is the capital of the European Union). Again it is the study of neuropsychology, of the effects of brain damage, that provides strong support for these distinctions, with some patients showing, for instance, intact ability to perform old motor skills and the unimpaired capacity to learn new ones despite suffering from an amnesia so dense as to be incapable of remembering who they met just moments before. Exactly how these different memory forms may be related to one another, and just how secure some of these distinctions are, are the subject of much ongoing research by cognitive scientists. This is no place to go into the details (though looking at the book by the English psychologist Alan Baddeley listed at the end of the chapter will give the interested reader a way into the technical literature). But what is clear is that memory mechanisms can no longer be thought of as all one single thing. And no amount of astute observation of self and others in the normal course of everyday life would reveal these differences, and certainly not with the clarity of experimental studies. The moral of the story is that folk psychology is something quite different from the picture of the mind that scientific psychology is slowly revealing.

The example of different memory mechanisms gives us the answer to your second question: 'How certain can one be that a hypothesized mechanism does actually exist?' Well, of course, in a sense, nothing is certain in science. Newtonian physics reigned supreme for hundreds of years, and remains an extremely accurate predictor of things like planetary movement. Yet 20th-century physics overturned all previous theory in physics, so one must assume that our understanding of physics may change yet again in this century – indeed, there are indications that this will happen. And if physicists can change their minds about the correctness and accuracy of their theories, who would bet against our theories about the mind/brain altering, and altering in a big way, as novel empirical methods are developed and fresh theoretical insights arise. For example, it is extraordinary that psychology came to realize the huge importance of the human ability to understand that others have intentional mental states, so-called Theory of Mind (see later in this chapter), only about 20 years ago. How could we have studied the mind within a formal, scientific framework, much assisted by the attentions of philosophers, for 130 years and not realized the importance of this ability – indeed, that it existed? And more than just Theory of Mind, our understanding of infant cognition in the round has changed dramatically in the last few decades and forced an entirely different view of the development of knowledge and understanding in children from that held prior to the 1980s. Given the general and deep uncertainty that underlies our knowledge about how the mind and brain work, as indicated in chapter 1, it is a safe bet that a great deal will change in the coming years. In the end all we can do is hold on to our seats and watch the science unfold, as it will surely do. Yet as is clear from what has come before in this book, and will be reinforced by what comes later in this chapter, the history of scientific psychology is one of increasing, if slow, accumulation of knowledge about how the mind/brain is structured and how it works. Prior to the cognitive revolution in the 1950s and 1960s, we had very little understanding of memory at all. Hebb's hypothesis of different mechanisms for different forms of memory came largely from a consideration of what neurological events might underlie memory. Psychological experiments subsequently confirmed his general hypothesis, and greatly increased the richness and scope of our theories of memory, which doubtless will change further with time. But that is the nature of science. If, despite extensive experimental and observational study, our ideas about the mechanisms of the mind did not change, then that really would be cause for concern. In that event, we would not be doing science.

It is worth noting what has been implicit in the last few pages, which is that psychological theory is being developed and tested by an increasingly strong array of empirical methods. Once the cognitive revolution freed psychology from the intellectual calamity of behaviourism and allowed it to hypothesize the existence of 'hidden' causes, causes which cannot be directly observed (behaviourists demanded causes that could be literally seen), it not only matured into a science alongside all other natural sciences that cannot directly observe causal forces (nobody has ever literally seen the inside of an atom or placed their fingers on a gene), but it also developed a wealth of new methodologies. Powerful experimental methods developed for use in adults began to be applied to ever-younger children, and by the 1980s data on infants were being generated that had previously been completely beyond the reach of developmentalists. The rise of cognitivism was not confined to the study of humans. Primatologists increasingly looked at their subjects as cognitive agents rather than just as behaving animals. This allowed ever more imaginative experiments to be carried out and direct comparisons with humans to be made. Furthermore, while the study of brain-damaged patients had been a fertile source of information about the structure and function of the mind/brain since the middle of the 19th century, neuropsychology was hugely reinforced from around the middle 1980s by the development of an array of imaging methods which allow the distribution of activity in the brain (measured, for example, by changes in regional blood flow and energy consumption) to be observed in normal, conscious people (and animals) carrying out various tasks. Cognitive neuroscience, comparative cognitive studies, developmental approaches and clever experimental cognitive psychology combine with the ability to simulate events in the brain using computer models of neural networks to form a powerful combination of methodologies for understanding the mechanisms of mind.

As to your first question, about what form a psychological mechanism takes, this leads us into much more difficult and fraught territory. Most, but by no means all, cognitive psychologists subscribe to what philosophers of mind call the functionalist school. Functionalism considers that mental mechanisms and mental events are to be understood and recognized in terms of their causal roles within the total functioning of the mind as an integrated system. So, just as, say, the role of the distributor in a car engine is to provide the electrical input to each cylinder which produces the spark that ignites the air–fuel mixture which then explosively drives the piston up and down, so working memory is thought to have a specific function. This is believed to be to act as a temporary store for outputs from both sensory surfaces and their associated buffer memories, and also from intermediate computational mechanisms during problem-solving, and to perform further computations on these temporary outputs before passing the result as an input to some other cognitive mechanism, perhaps one of the systems that computes a motor response. Another hypothesized mechanism is some kind of supervisory attentional system which sorts and prioritizes the outputs from information computed in parallel by sensory channels, in effect providing a kind of executive control of multiple, and sometimes conflicting, information. Thus, functional mechanisms are not necessarily identified by their material properties, but by what they do and cause to happen.

It is the 'not necessarily' of the previous sentence that leads to problems. One of the important features of functionalism is that it allows psychologists to get on and study the mind without having to wait for advances in brain science – indeed, some see it as freeing psychology altogether from knowledge of the brain. And that is a very big advantage of the approach. But this then leads to argument as to whether other complex systems that perform computations on information that leads to actions and problem-solving, such as computers or whatever the brains of Martians are made of, also have minds. Well, that is not our problem here. Artificial and alien intelligences aside, what is certain is that the human mind is most certainly one with the human brain. To the purist functionalist, that is an irrelevance. However, while it might offend the sensitivities of philosophers of mind to try to match the knowledge that is coming from cognitive psychologists working within a functionalist framework with the knowledge that is coming from physiology, neurochemistry and various forms of brain-imaging, to mention just some branches of neuroscience, such matching is exactly what the sciences of the mind and brain should be doing. Cognitive psychology feeds vital information to neuroscientists about what they should be looking for, and neuroscientists in turn give psychologists clues as to how to sharpen up their picture of the mind. For example, a recent brain-imaging study suggests that a part of the brain long known to be involved in language processing, Wernicke's area, is also activated when people are presented with predictable sequences of events, even though they are not consciously aware that they are detecting the predictability. The significance of this is not yet understood, but given that language has rule-governed structure which offers a substrate for prediction, this is a finding that is likely to feed into the further development of theories about the evolution of language and how exactly it develops in infants and children.

To summarize this first set of points, then: most psychologists think of the mechanisms of the mind in causal terms, in terms of what they do. We cannot be certain that the mechanisms postulated really do exist, but cognitive science advances through the interplay of an increasingly powerful empiricism and theory development, just as do all other sciences, and the mechanisms currently on offer are the outcome of the same processes of science as operate in other disciplines. Finally, psychological science is an entirely different enterprise from the 'natural' or folk psychology that all of us practise. The latter is a consequence of our sociality and comes naturally to every person, in much the same way as, say, our ability to construct a visually coherent scene from the input streaming up a million nerve fibres from each of our eyes. We don't think about it – we just do it. Science, on the other hand, is an effortful, difficult process that leads, as we noted in chapter 1, to unexpected insights into the world and a form of knowledge that departs widely from the quotidian. In the same way as the sciences of physics and chemistry describe worlds not available to ordinary experience, so psychological and other cognitive sciences are revealing structures and mechanisms that are not directly available to us as we go about our daily lives.

This brings us to the second point that needs to be cleared up before we can turn to the mechanisms of intelligence themselves. The human capacity for creating and entering into culture embraces all psychological processes, mechanisms and functions. For example, imitating an action, which at least at some point in human history was a very important part of culture, involves the translation of visual input, including the construction of a coherent visual scene as mentioned in the previous paragraph, into a matching motor output. Yet we could not live within the world of visual sensation without using that ability for scene construction, and we use it constantly when awake and with our eyes open. In other words, constructing a visual scene may be essential for imitation, but it is also essential in all other visual activities that do not involve imitation. Similarly, language, the kingpin of culture, involves attending and being sensitive to auditory information, which is something we do all our conscious lives, even in non-linguistic environments. We use our memory and perceptual systems both in and out of culture. And many of these mechanisms and processes, such as visual-scene construction and the sorting of auditory input, are not human-specific – indeed, they are not ape- or primate-specific. So while culture may employ virtually all of our psychological processes and mechanisms, meaning they are all essential to us as creatures of culture, there must be certain mechanisms that are what makes the difference between the kind of cultural animal we are and the kind of cultural animal chimpanzees are – and, indeed, the kind of social but non-cultural animal that, say, zebras or elephants are. If, then, we are looking for the essential mechanisms of culture, we are concerned with what makes us different from other animals, and especially from chimpanzees, rather than all those supporting mechanisms that we use in any and every activity and share with so many other creatures.

What those mechanisms may be

We can be certain that some mechanisms are human-specific and essential for culture. A second group of mechanisms may not be human-specific, or at any rate the empirical jury is still out as to their uniqueness to our species, but they probably have played a crucial role in the evolution of human culture. Language is in the first category. It is unique to our species and at the heart of human culture. There are no doubts concerning language as a mechanism, or more probably a set of mechanisms, that is causal in human culture. Theory of Mind and susceptibility to social force may also be unique to humans. The second category comprises mechanisms that underlie imitation and the formation of what will be referred to as higher-order knowledge structures, which may well be present in chimpanzees, although they have not yet been sought for, and are certainly important in culture. Doubtless most, if not all, of these are very closely interlinked, especially during infancy and early childhood enculturation, to the point where their interdependence may cast serious doubt on their being separate mechanisms in the strictest sense of the word *separate*. It is for convenience of presentation that they are given separate headings in what follows.

A final few words of summary to help the reader navigate through what is becoming a long argument extending across all the chapters of this book. My basic aim, remember, is to explore ways of marrying the biological and social sciences (chapter 1). The principal vehicle for doing this is the conception of culture as an expression of human intelligence, and that brings into sharp focus the implications of the evolution of intelligence (chapter 2). This all positions the argument firmly within the school of ideational theories of culture (chapter 3). To paraphrase Goodenough: culture is what people have in their minds, how they acquire knowledge and beliefs, and how they transmit these to others. Chapter 4 considered process approaches and stressed, as above in this chapter, how processes must be embodied in causal mechanisms. Which brings us to what we are doing here – exploring a choice of mechanisms guided by that ideational approach to culture. We are looking for mechanisms whereby knowledge and beliefs are gained and transmitted.

Concepts, schemata and other higher-order knowledge structures

In the previous chapter the example was given of a meme that had to do with all the members of a household of employable age being expected to contribute to the costs of the enterprise – what was termed the I-expect-you-to-contribute-to-the-costs-of-running-this-household meme. One of the main points of the example was to illustrate just how much knowledge is wrapped up in a meme of this kind, such as what a cost is in terms of money, the expectation of sharing some of one's income based on a belief in fairness and reciprocity, and the implication that everyone should be out there earning some money if they can. And each of these things in turn hinges on understanding what, for example, money is and what is and is not fair. OK, this is, admittedly, a deliberately complex choice of meme. But when you are told that the *Délicieux* is the best-value eating place in town – a simple, commonplace meme of social interaction – you need to know what a restaurant is and be able to understand the concept of good value in order to have any idea at all what the information being imparted to you means. And when I tell you where this place is, the information is of no use unless you know the area in question and how to get there – and if you don't know that, you had better be able to read a map, but what is a map?

Now these are not exceptional instances of the kind of thing we exchange culturally. There are much more complex memes, such as democracy, relativity theory and moral imperatives. Here, then, is a challenge to you, dear reader. Can you think of any memes at all, apart from simple motor acts, passed between people older than about three or four years of age which do not encompass a great deal of prior knowledge? Should you be able to, I would, as they say, like to hear from you. The point is clear. Apart from early infancy, when little or no culturally specific knowledge has yet been acquired, and leaving imitation aside, nothing that we transmit to each other occurs in the absence of such background knowledge. Even when I give someone my home telephone number (which is as simple as information gets) it is based on the assumption that the recipient knows both what a telephone and a home is, and hence why the number is privileged information. This background knowledge is what I refer to here as a higher-order knowledge structure. Structures of this kind are the substrate of all meaning.

High-order knowledge is a part of semantic memory, and important to semantic memory is how words come to have meaning (which is an example of the arbitrary distinction between different interconnected mechanisms, in this case those of memory and language). Furthermore, as words come to have meaning, they form clusters, or higher-order meanings, which we call concepts. The importance of concepts is that they are the basis for making sense of the world and communicating that sense with others – they are at the root of propositional thought. Animate objects, for instance, are a very important subset of objects – they are the things that we eat and which eat us. It is not surprising, therefore, that children rapidly form explicit meanings for the names of animals. A bird is one thing, and different from a dog, which is another. But exactly what is a bird? Well, it has wings and feathers and it sings and flies – although not always, of course. Ostriches and penguins neither sing nor fly. And we know from careful experiments that when people categorize creatures like penguins they take longer over doing so than when categorizing a blackbird precisely because the former violate the common assumptions about what birds typically are and do, which the latter exemplifies. Opposing theories of just how a concept is formed have yet to be settled. Some postulate the existence of models made up of necessary and jointly sufficient attributes, an approach sometimes referred to as the classic theory; others subscribe to the more relaxed Wittgensteinian notion of family resemblance, in accordance with which a rather fuzzy exemplar bearing the common attributes of the category is built up – so-called prototype theory. The evidence currently favours the second of these. But no matter, for our purposes. What does matter is that while we slowly establish separate concepts of bird and dog, we also bundle them together into higher-order concepts like pets or animals, which overlap but are separate. In turn, the animal concept is different from the plant concept, but they combine to form the living-forms concept, which includes bacteria and fungi. And to complicate things, fungi and worms might form a creepy-things concept, while sable antelope and tigers might be part of a noble-beast concept.

Two things immediately become clear. The first is that semantic memory in the form of the concepts that words exemplify has a structure. Cognitivists argue, as they do about most things, as to the nature of that structure. At first it seemed the structure must be hierarchical, with birds and dogs subsumed under the higher level of land animals or vertebrates, which in turn was subsumed under all animals, and so on. Experiments on how people do relate different concepts, however, led to suggestions of less rigid network structures for concepts. In part this was because of the second point that the examples above make, which is how flexible concept formation is. There is nothing rigid about a mechanism that can take a class of thing, such as a dog, and place it simultaneously within a number of different concepts (a companion, a domesticated working animal, a carnivore, a mammal, and a nuisance or danger if you earn your living by delivering things to people's homes). How one understands the world and communicates that understanding depends on which concept of dog, bird, fork, bicycle or voting system one is employing at any one time. Insofar as culture is transmitted knowledge and belief, concepts are very important packages of that knowledge and those beliefs.

The fluid yet focused forms of knowledge represented by concepts are not the only kind of higher-order semantic memory. In the previous chapter mention was made of Bartlett's studies of memory for narrative. Bartlett revolutionized thinking about memory. Our memories, he argued, are not passive re-membering of stored detail, but acts of creative reconstruction. Memories change because each of us reworks and reinterprets our experiences. The reworking is determined partly by how we would have liked things to have happened, and partly as a result of the effects of unconscious generic knowledge structures that form anchoring points round which we create memories. He called these knowledge structures, which are complex models of the world we live in, schemata. (The notion of the importance of schemata was independently developed by the Swiss psychologist Jean Piaget, for whom a schema was an inferred element of cognitive order that determined the recurrence of actions and, as cognitive development advanced, the organization of higher-order schemata. While Piaget's theory of cognition was ground breaking, it contains features which have caused it in recent decades to rather languish in the backwaters of cognitive psychology. This may yet change, but the renaissance of schema theory in the last couple of decades owes much to Bartlett's formulation and not much to Piaget's. So we will confine ourselves to Bartlett.)

Schemata encompass all aspects of experience and activity, and form, if a metaphor from astronomy is not too way out, mental centres of gravity which attract to themselves particular experiences which they then shape. So, to use one of Bartlett's own examples, in many cultures each person has a schema for the office, which is a place of work-related activity, usually focused on a single kind of enterprise, equipped with tables, chairs and appropriate technology (typewriters and telephones in Bartlett's time, computers, printers and fax machines today). The office schema does not include machine tools, swimming pools or people in police uniform. Constructing a memory for an event or story relating to an office is drawn into the gravitational pull of the office schema, which affects the recall of the event or story. Details of the original that are not usually a part of the schema, such as the presence of a power drill, are forgotten (unless they are so remarkable, as a swimming pool at the office would be, as to be a spur to memory), while items might be added that are not part of the narrative but are prominent in the office schema (the presence of desks, for instance). Thus it was that in experiments on the memory for folk tales from unfamiliar cultures, Bartlett found that memories were systematically degraded by subjects who were creating them on the basis of schemata that were inadequate for the task of remembering the material concerned.

Bartlett's ideas were ahead of his time and did not chime well with the prevailing atomistic, associative approaches to memory that marked the psychology of the 1940s and 1950s. Then came the cognitive revolution and with it the conceptual freedom to postulate unobservable mental mechanisms. The 1970s also saw the rise of artificial intelligence. Some reasoned that in order to build intelligent machines one should model them on the highest natural intelligence on Earth. Marvin Minsky, an American computer scientist, realized that one of the characteristics of human intelligence is the ability to form generic knowledge of just the kind Bartlett had written about 40 years previously. Minsky adopted the notion of schemata, which he called frames, and developed them in formal ways to fashion them for use in computer programmes. So the frame for schools, for example, comprised a cluster of features, such as the number of classrooms, desks and teachers, to each of which could be attached specific values assigned by experience.

David Rumelhart, an American psychologist, and his colleagues reworked the frame notion back into psychological form. Schemata, they asserted, have four major characteristics. They have variables, they can embed one within the other, they represent knowledge at all levels of abstraction, and it is knowledge they represent, not just definitions. By this they meant schemata are based on real-world experience, on networks of interrelations, actions and goals. In Rumelhart's own words, schemata 'represent knowledge at all levels – from ideologies and cultural truths to knowledge about what constitutes an appropriate sentence in our language to knowledge about the meaning of a particular word . . .' Above all, schemata are active, dynamic and generative structures. Subsequently, Schank and Abelson extended this kind of thinking to generic knowledge of actions in specific settings – for example, what people normally do when entering a restaurant – which they called scripts. At Yale University, Schank further developed his ideas about complex memory in terms of memory and thematic organizational packets – MOPs and TOPs.

Now these were much more than just exercises by people fooling around on computers and creating theories with amusing acronyms. Schema theory finds fruitful application in many areas. One illustration of this is the fascinating work of Katherine Nelson, at City University of New York, on the development of schemata. Nelson's studies are among the most detailed accounts of how children of different ages acquire knowledge about their worlds. One of the more interesting things they reveal is that the cognitive competence of young children is greater than indicated by more formal and abstract tests of children's thinking of the kind that Piagetian developmentalists routinely administered. They take the study of what children know into the real world of children's everyday life. Another application, of interest to anthropologists, has been the work of David Rubin, who has applied schema and script theory to understanding memory of oral traditions. Before the invention of written script just a few thousand years ago (and until very recently writing has been available to only a privileged minority) the only vehicle for directly transmitting culture across generations, apart from the imitation of manual skills, was the oral tradition of story-telling. Schema theory, together with insights into the role of rhythm and song, provides us with the beginnings of understanding of how preliterate cultures, which occupied so long a period of human history, worked, and how they were constrained and shaped by the mechanisms of human memory because, by and large, memory was all they had with which to maintain complex cultural rituals and beliefs.

Are higher-order knowledge structures present in other species? Famous experiments at Harvard University in the 1970s on pigeons indicated that these birds are able to extract common features, which is the basis of forming a concept, from an array of inputs. (Most who write of this work refer to 'category formation' rather than concept formation.) The pigeons were presented with 40 different slides that had a single common feature, which was the image of a tree or trees, though the trees differed in type and orientation. They were also presented with 40 slides without images of trees. The birds were rewarded only if they pecked at images containing trees, not if they pecked at images without trees. After many hundreds of trials the pigeons were able to discriminate accurately between scenes with and without trees. Importantly, the ability to make the discrimination was maintained on test trials, in which the pigeons were presented with slides they hadn't seen before, which rules out any interpretation of the results in terms of memorizing individual images. The pigeons had acquired some general knowledge about the visuo-spatial configuration of trees. Subsequent experiments extended such findings to other categories, including fish and people. Similar findings with different species of monkey have also been reported, as well as a positive study using a chimpanzee. None of this came as a surprise, at least in relation to the monkeys. It had been known for decades beforehand that monkeys could solve problems such as oddity and so-called matching-to-sample tasks, in which they are able to learn, say, that it is always the different element accompanying a cluster of otherwise identical elements that must be responded to, even when they are presented with shapes they have never before encountered. And the fact that chimpanzees fashion simple tools to 'fish' for termites in nests differently situated and with different external configurations suggests they can and do categorize objects like termite nests.

At present we know not much more than this about other animals, but one important difference between chimpanzees and humans may be the degree of higher-order that each can categorize. The ability to abstract to the degree of categorizing both chess and football under the concept of games is a very high order of abstraction indeed, and unlikely to be available to non-human apes. This is, however, an empirical question waiting for an answer, though nothing in that *Nature* paper describing culture in chimpanzees suggests that what is transmitted between members of a social group is any kind of higher-order knowledge. This is in contrast to humans, among whom the great majority of culture information exchange takes this form.

A final word on higher-order knowledge structures: causality does not run just one way between such psychological mechanisms and the wider culture of what is shared. The creation by each child of its own knowledge is also a product of intense and prolonged interaction with others, with peers who are actively engaged in establishing their own knowledge structures, and with carers and other significant people who already carry in their minds and behaviours the knowledge and beliefs that characterize the culture. This two-way street of influence between psychological mechanism and culture, between the biological and the social, is not unique to the formation of schemata and scripts. It is the primary characteristic of all the psychological mechanisms that are essential to the existence of culture. It is yet another example of the complex causal architecture that characterizes all of biology, and hence all of the social sciences too.

Imitation

Reference has been made several times so far in this book to the culture, or protoculture, of songbirds and chimpanzees. Careful observation (in the case of chimpanzees) and experimentation (with songbirds) has shown with certainty that these animals live in communities, many, if not all, members of which share certain behaviours, such as gestures and tool usage among chimps, and, of course, song among songbirds. Such shared behaviours are the result of learning from others what those others themselves acquired through an act of learning. This, it has been argued, is the core quality of culture. But to ask again that key question: is chaffinch or chimpanzee culture the same kind of culture as we humans have, or is it different, and if different, different in what way and for what reasons? Most people would answer intuitively that human culture is massively and manifestly different from that of any other species, as would I. But intuition cuts no ice in science. Really to answer the question we have to look at the detail, and the detail lies in the mechanisms.

There are, broadly speaking, two ways in which one individual can learn from another. The first is indirect: some aspect of the behaviour of one individual results in the other individual adopting similar behaviour because some feature of the world has been revealed to them. Take the example of someone coming to stay as a guest with me in my corner of north London. I walk them round the local streets, and because we happen to be near a particular shop and are short of milk, we go in and make a purchase. Thereafter, whenever my guest needs milk, they go to the same shop. I have done no more than inadvertently draw their attention to the shop and the fact that it stocks milk. I have not explained in so many words that this is the place to purchase milk in preference to some other, perhaps more distant, shop, or even that we need milk in the first place. Compare this with the second way in which my guest could learn something from me. This is the direct transmission of information between myself and my guest, when they ask where they can buy some milk and I reply that they must turn right on leaving the house, left at the corner and up one block to Georgeou's store on the right, which sells milk. Both are forms of what has come to be called social learning, but they differ in that crucial respect which, for the moment, we can call the difference between direct and indirect transmission of information.

The differences, substantial or trivial, between different forms of social learning have become an academic growth industry in recent years. So, too, has the distribution of different kinds of social learning in different species. And it is these differences that account for the unique nature of human culture. First let's consider a classic case from the animal behaviour literature. Although now a diminishing feature of British culture, before the establishment of supermarkets the delivery of milk to the front doorstep of people's houses was a widespread custom. In the late 1940s the design of milk bottles changed, a top seal being introduced that was easily penetrated by the beak of a small bird. Within a remarkably short time, the puncturing of milk-bottle tops by tits had become widespread in the United Kingdom. At first, it was assumed that following the discovery by an original smart tit that the bottle tops could be pierced, other tits observed and then imitated this behaviour, and were in turn imitated by yet others – and so the behaviour spread rapidly. Well, it was a nice story. But alas, 20 years later experiments showed that when one tit observed another piercing a bottle top, what really happened was the attention of the observer was drawn to the seal itself. Thereafter, penetration of the seal by the second tit came about through that bird's own trial-and-error learning, not direct imitation. Stimulus or local enhancement is the technical term for this particular kind of indirect social learning. Well, given that tits ended up with common, if indirectly shared, knowledge about milk-bottle tops, does this mean that the common tit, a perfectly ordinary European bird, has culture? Well, I don't see why one should rule against this, but the culture is very limited, and it is the result of a specific mechanism of attention combined with individual learning. This is not like our culture, because the transmission of information is indirect. But when I look at you serving badly on a tennis court, then step up to the base line and say, 'No, you're doing it wrong. Watch me and then do exactly as I do', and you do, then you are imitating me. You are 'learning to do an act from seeing it done', in the words of the original definition of imitation by the American psychologist Edward L. Thorndike. As far as is known, tits can't do this. (Of course, imitation of complex acts is seldom a case of one-trial learning. Anyone who has learned or taught a skill like playing tennis knows it can be a slow and difficult process.)

Now imitation has always occupied a special place in almost everyone's story of the evolution of our species. The reason is simple to understand. The oldest stone tools date back to over two million years ago – indeed, the very oldest, from the Hadar region of Ethiopia, may be as old as two-and-a-half million years. Certainly by two million years before the present, stone tools become a permanent feature of human palaeoarchaeology. *Homo habilis*, the first species of the genus *Homo*, was certainly a tool maker and user, and it is possible that its ape precursors were too. The consistent appearance of tools over tens of thousands of generations of early humans, together with the undoubted modern human ability to imitate, goes against the notion that tools were independently and repeatedly invented by each generation. In the light of contemporary human imitative abilities, it also seems unlikely that the long history of human tool technology was supported by some indirect mechanism of learning of the kind used by tits to conquer the British milk bottle. And yet at some point in human prehistory our ancestors must have had only the cognitive abilities of the Miocene ape that was the common ancestor of humans and chimpanzees, and that ancestral ape was probably, like chimpanzees, a tool user (as are certain species of corvid bird), yet may have lacked the cognitive mechanisms necessary for imitation. This may sound paradoxical, but whether modern apes can imitate remains unresolved. This is a highly contentious issue among psychologists and primatologists.

What looks like imitation by both chimpanzees and orang-utans has been recorded in a number of vivid accounts. The evidence from the wild-living chimps described in that *Nature* paper apart, the primatologist Frans de Waal has given a fascinating account of the behaviour of captive chimps in a recent commentary in the journal *Behavioral and Brain Sciences*, which ran an article about imitation. De Waal describes how sometimes the infant chimpanzees in the colony at the Yerkes Primate Centre get their fingers stuck in the fencing of the compound, and because forceful pulling only hurts them, the adults have learned not to apply force when trying to free them. During these tense incidents the whole colony becomes agitated, and it takes some time before the infants manage to extricate themselves by chance. On several occasions de Waal has seen other chimps mimic the plight of the infants. For instance, 'One older juvenile came over to reconstruct the event. Looking me in the eyes, she inserted her finger into the mesh, slowly and deliberately hooked it around, and then pulled as if she, too, had gotten caught. Then two other juveniles did the same at a different location, pushing each other aside to get their fingers in the same tight spot they had selected for this game.'

De Waal's comments were a response to the main article in the journal, by the primatologists and psychologists Richard Byrne and Anne Russon, who described the behaviour of a rehabilitated colony of orang-utans in Indonesia, which included 'siphoning fuel from a drum into a jerrican, sweeping and weeding paths, mixing ingredients for pancakes, tying up hammocks and riding in them, and washing dishes or laundry'. The descriptions are fascinating, but unfortunately there can be no certainty that what these animals were doing was imitating in the same way that a human learns to imitate the action of serving a tennis ball. There are several reasons for this uncertainty. One is that many such stories (though not necessarily those just cited) are anecdotes. Now anecdotes in science do not have high status as data. Some are the result of one-off observations, and replication, especially in areas of biology as intrinsically unclear and complex as primate behaviour, is generally held to be essential for any degree of confidence that what is being reported is really true. Anecdotes, by their very nature, are also subject to creative adornments by the teller, albeit inadvertently, and sometimes involve animals of unknown history and which may have been specifically trained by humans to do what they are reported as doing. Repeated anecdotes, preferably from different observers, that tell the same story certainly do improve confidence in the validity of what they report. But what one really wants is some kind of permanent record, such as film or video, which can be minutely and repeatedly examined. And when such records are available, they often demonstrate a very high degree of variability in behaviour, which raises the question as to whether what is being seen is real imitation or something else. Matters are not helped by clear cases of human imitation of complex acts, such as serving a tennis ball, also showing wide variation between the model and the imitator. Furthermore, anyone with experience of teaching a skill knows that imitation seldom occurs in isolation of other forms of learning. In a tennis lesson, for example, there is constant vocal correction and instruction, both linguistic and non-linguistic; there may be actual physical positioning of all or part of the body of the learner; and a great deal of indirect, attentionally guided information is brought to bear by the imitator, such as it is the ball that is tossed in the air and struck by the racket, which is held in the hand, not the other way round. (The latter might sound silly, but it is just this kind of detail that makes attempts to simulate tasks such as imitation using artificial intelligence so difficult.)

Something that can look like imitation but isn't has been called emulation learning by primatologist and human cognitive developmentalist Michael Tomasello. Tomasello gives as a credible example a mother chimpanzee, observed by its infant, rolling over a log and thus revealing the insects under it. Edible insects sheltering beneath a log is what the young animal learns, not how to roll a log over, which it either already knows how to do, or will learn to do on its own and not as an act of imitation. This is an instance of the young animal learning what is technically referred to as the affordances of objects – what can be done with them, what they can do, and in this case what they conceal, but not exactly how to do the doing. In a similar way, when one animal sees another cracking open a nut with a stone, what it might be learning is not the exact action of striking the nut with the stone, but that stones and nuts have these affordances (as do balls and rackets), which it then learns to exploit on its own. (Tomasello is at pains to point out that emulation learning is a pretty smart form of cognition. It is not necessarily inferior to imitation, just different.)

Emulation learning is not the only possible explanation of what may superficially look like imitation. Convergence on a favoured form of behaviour may simply arise because it is the most 'obvious' solution to a commonly perceived problem. A variation on this theme is a kind of mutual shaping of behaviours by infant and caretaker within a confined 'behavioural space' leading to similar behaviours in different individuals. 'Ontogenetic ritualization' is the term used by Tomasello to refer to such similarities, an example of which is the raising of outstretched arms over the head by young infants as a signal that they want to be picked up. Many human children do this, of course, but the behaviour is not necessarily acquired by imitation. Indeed, it probably is not because it is often observed in children who have had no opportunity of observing it in others.

Well, with such alternative possible explanations, the only way of sorting them out is to do experiments. Tomasello did just that, comparing chimpanzees and two-year-old human infants who were shown how to retrieve an out-of-reach object with a rake-like tool. Half of the subjects of each species observed one (efficient) method, and the other half a second (inefficient) method. The result was that the human children almost always did what they had seen, copying both inefficient and efficient methods equally often, whereas the chimpanzees did lots of different things to retrieve the object, irrespective of what they had seen done. In some cases the emulation learning of the chimps was more successful than the imitation of the inefficient method by the children, who persisted in this less successful, imitated, adult behaviour.

Despite the compelling result of this experiment, some argue that there is not yet sufficient evidence to be certain one way or the other on imitation in non-human apes. Tomasello is certain, though, that apes do not imitate, and his certainty stems from the conviction that non-human apes are unable to put themselves in the intentional shoes of others – they cannot understand the goal that the observed demonstrator has in mind when performing an act – and such mind-reading involves another mechanism, a Theory of Mind (to be described in a later section), which non-humans just don't have. That a Theory of Mind mechanism is essential for imitation, and that animals like chimpanzees do not have it, remain at present contested issues. But it is just such disagreement that makes this corner of science so fascinating.

Another way of looking at imitation was proposed by Byrne and Russon. They make a distinction between 'action level' imitation, which is Thorndike's learning to do an act from seeing it done, and 'programme level' imitation. Much behaviour is hierarchically organized within an overall programme of actions, with each level of action guided by subgoals which serve as steps towards the final goal. For example, if my overall goal is to eat a plateful of eggs and bacon, I achieve that by, first, taking the ingredients out of the refrigerator, second, gathering various utensils from cupboards and drawers, third, turning the cooker on, fourth, placing the bacon in a pan, fifth, frying it to the required degree, and so on, until I am finally ready to eat. Byrne and Russon suggest that some sequences of ape behaviour, such as their simpler acts of food preparation, should be viewed in this way, and that what these animals are capable of is copying the overall sequence of acts, including subgoals, without the necessary accompaniment of 'action' imitation within each level of the programme. While this departs from the narrower definition of imitation that most people focus on, it does seem a reasonable alternative explanation for what often looks like imitation by apes. It does not, though, solve the puzzle of action imitation.

Before briefly addressing the matter of how action, or Thorndikean, imitation might work, we need to consider those songbirds again. Nobody now doubts that such birds must hear conspecific song in order to reproduce it, and that the reproduction is astonishingly accurate, down to the acquisition of regional dialects – they have local accents, just like humans. How can small-brained creatures lacking the mass of cerebral cortex that apes have achieve a cognitive feat that is doubtful in our nearest living relative? Part of the answer is that brains don't have to be big, or structurally similar to ours, in order to be capable of highly specific forms of learning – witness the abilities of honey-bees to gain positional information about food sources by way of the dance of conspecifics and then to translate that, with a correction for the changing angle of sunlight with time, into accurate and efficient foraging behaviour. And that is the point about birdsong. It is a highly constrained form of learning (see chapter 2) triggered by the very specific developmental event of hearing the song of con-specifics. We, on the other hand, can imitate whole body movements, facial expressions, gestures, speech sounds and non-speech vocaliza-tions, among other things. With hundreds of millions of years of evolutionary time separating us from birds, it is reasonably certain that the mechanisms songbirds use are different from those we humans use.

So what might the mechanisms we use be? 'Learning to do an act from seeing it done' requires some kind of perceptual-motor device by which visual input is matched in some way to a motor programme. The Italian neurophysiologist Giacomo Rizzolatti and his colleagues have shown the existence of what they call mirror neurons in a specific part of the cortex of monkeys, which fire both when monkeys perform a movement and when they see that movement performed by others. Doubts about monkeys, never mind apes, being able to imitate are considerable, but these findings suggest some mechanism which, even if it is not capable of supporting imitation in monkeys, nonetheless has a component basic to imitation. This is the ability to recognize the match between the movements of another conveyed as visual input, and one's own movements and their sensory inputs, which must com-prise a sensori-motor programme built on sensations from receptors in joints and muscles. If such a mechanism is widespread in primates, or at least widespread in monkeys and apes, it may be the substrate evolution built upon and elaborated in the genus *Homo*. Recent brain-imaging studies by Rizzolatti and others have shown very specific areas of human cerebral cortex to be activated during the imitation of finger movements, thus supporting the idea that imitation may be the consequence of specifically evolved brain/mind mechanisms rather than of some general-purpose device associating sensory input with motor output.

Over the years a variety of theories of imitation have been offered. Among these have been a number of associationist accounts, which have in common the assumption that links, or associations, are established between the various forms of sensory input caused by the observed action and the sensory inputs caused by production of that action. Associative learning is very widespread, however, being an attribute of thousands of different species, and this poses a problem for such accounts of a form of learning that is possibly restricted to just one or a very few species. The position can be saved by arguing that what is special about imitation is the way that what is to be associated is bounded, concatenated or transformed, but then the association becomes trivial and what is important is the bounding, concatenation or transformation. This is a feature of what are referred to as transformational theories. Most such theories postulate some kind of supramodal representational system within which the transformed sensory inputs of the imitator, be they visual, auditory or of some other variety, can be matched with the transformed sensory inputs caused by the muscle and joint senses when the action is performed by the imitator. In other words, the motor programmes that control specific actions are represented as transformed sensory representations 'written in the same language' as the transformed sensory inputs from other senses, so the two can be brought into various degrees of fit or matching.

It is the very complexity of the computational requirements of transformational theories that makes the postulated mechanisms more likely explanations of imitation. This is because if imitation is indeed a phylogenetically rare cognitive skill, it may be so because it is based on a rare cognitive computational ability, not just the widely present capacity for associating sensory inputs with one another and with motor outputs.

Imitation receives prominence in the writings of those interested in cultural evolution because it is one way of keeping copying fidelity high (see chapter 4), and this, in turn, is necessary for what Tomasello calls the ratchet effect. The ratchet effect is the result of an interplay between faithful replication and the creative extension of acts and artefacts such that modifications are cumulative – nothing gets lost and many things get better. This is a phenomenon unique to humans.

But however important imitation might have been in human evolution, especially in the evolution of first culture and the subsequent evolution of contemporary culture, and however important it might be in the very early enculturation of every child, it nonetheless, in the context of modern culture, pales into insignificance when compared to another form of non-genetic transmission of information, which is what we turn to now.

Language

In a wonderful review of language the evolutionary psychologist Steven Pinker demonstrated the absurdity of a claim by the *Guinness Book of Records* that a monstrous sentence by William Faulkner of some 1300 words, beginning 'They both bore it as though in deliberate flagellant exaltation . . .', is the longest in the English language. By adding to it 'Faulkner wrote . . .', Pinker made the entire 1300-word sentence a mere phrase and increased the length of the sentence by two words. He further pointed out that anyone could then better his effort by writing 'Pinker wrote that Faulkner wrote that they both bore it . . .', a process that could go on for ever. There is no such thing as the longest sentence because, the limitations of memory notwithstanding, sentences can be infinitely long. What Pinker was pointing to is what the linguist Noam Chomsky refers to as 'discrete infinity', a property language shares with numbers, which also have no end. Any number, no matter how enormous, can be increased by adding to it. While we cannot be certain that no other animal has this understanding, it is probably another unique feature of human cognition.

The creativity of language – or generativity, to linguists – gives us a definition of language as the ability to generate an infinitely large number of messages from a finite, usually quite small, number of elements or symbols. It is this generativity that allows us to communicate with others about objects or events that are not present, about what was or might have been in the past, and about countless possibilities in the future, and all with a stunning degree of precision. We use language to invent scenarios that serve as a test-bed for actions, and we can do so by inventing entities and situations that do not exist and never have existed. In other words, we are the tellers of stories. The sheer quantity and quality of communication language gives us is not just unrivalled in any other species, nothing else comes near it. There have been persistent attempts over the last several decades to teach chimpanzees (and the occasional gorilla) language through a variety of means. The results have been uniformly negative in terms of both generativity and the appearance of rule-governed communication. (The former, in fact, derives from the latter.) Some of the apes trained have been exceptional compared to others, and some have acquired several hundred 'words' – or symbols, to be accurate. That, however, is not language, any more than is the ability of a 12–15-month-old infant to communicate with words. We can be charitable and use linguist Derek Bickerton's phrase 'protolanguage' to describe what apes and human infants have, which is the ability to produce one or a few symbols together but without rules governing their concatenation. But language itself, as a system of communication with rules of syntax and grammar that bestow meaning on symbols, is a human-specific cognitive trait that appears around 2 years of age. And, as pointed out in chapter 3, like culture, language is really a supertrait composed of a number of individual features, some of which, like working memory, have functions wider than those that just contribute to linguistic skill.

As we know from the previous section, as a form of extragenetic transmission of information language is not unique, and it probably was not the first to have evolved either. Given the phylogenetic distribution of indirect and direct forms of non-linguistic social transmission – not just imitation and other forms of social learning but grunts, cries, gestures and other movements signalling attention or emotion – it is almost certainly the case that these evolved before language did. And that is the only near certainty we have about the evolution of language. We have no idea when it first appeared in our species. Opinions range from two or more million years ago to just 100,000 years before the present, such disagreement deriving in part from definitions. Protolanguage probably does go back way beyond a hundred millennia. We also do not know why or how language evolved. It is tempting to go further and to say that, lacking a time machine that can take us into the past, we can never know why or how language evolved. No amount of archaeological artefacts or quantity of fossil evidence combined with reasoning based on what we know about language development in children or language impairment following brain damage in modern humans will produce a scientific account of the evolution of language. This is a story that can be, and has been, told a hundred different ways. So, with one exception, this book embraces the 1866 ban by the *Société Linguistique de Paris* on speculation about the origins of human language. If the tedium of speculative accounts prompted such drastic action by the French in the 19th century, all the more reason to resurrect the ban now after the last 20 years of fruitless conjecture that cannot be refuted. For those with a taste for speculation, however, I list at the end of this chapter a recent book by Bickerton and Calvin that presents a possible scenario for language evolution that is a great deal more credible than most offerings.

The single exception I want briefly to touch on is the role of gesture in language. One of the astonishing things we know about language is that it is a communication system that is not tied absolutely to ears and mouths. When hearing is absent, as in the case of deaf children, hands and eyes take over. One of the most notable features of language generally is that its development is invariant, no matter what specific language an infant acquires – which, of course, is the native language of its caretakers. Whether this be Japanese, English or any of the other thousands of documented languages of the world, every infant's babbling is the same. Recent research indicates that all spoken languages are based on four basic patterns of sound production by the human tongue, mouth and vocal chords. Gradually the babbling homes in on speech sounds characteristic of the specific language environment, and around a year after birth the first words are spoken. After a further six to eight months characteristic two-word utterances appear. The move from protolanguage to full language happens at about 24 months of age, and is followed by an explosion of syntactic complexity and the acquisition of new words. Chomsky notes that at their peak children may acquire new words at a rate of one per hour. And, very importantly, this all happens within an unstructured environment without formal tuition. Parents talk to their children, but they do not give them lessons either in grammar or in new words and their meanings. Indeed, the evidence indicates that most parents let the majority of grammatical errors go uncorrected.

Now, children who are profoundly deaf and raised by parents who use hand signs in place of spoken words go through exactly the same sequence of language-acquisition stages, and at roughly the same rate. Furthermore, brain-imaging studies have shown that the same areas of the brain as are activated in people with normal hearing when they listen to spoken language are activated in both deaf and hearing individuals expert at sign language when they read signs. There is more right-hemisphere activation in response to signs than to speech, and that does need explaining, but the important thing to note is that those parts of the brain that respond to spoken language are the same as those that respond to sign language – and they are not in the visual areas. It is also known that language impairments resulting from damage to specific parts of the brain are the same in people who speak through their mouths as in those who use their hands.

There is another interesting parallel between spoken and sign language. When people who speak different languages are thrown together as adults and have to communicate with each other, they do a lot of gesturing and miming, but they also develop a protolanguage called pidgin. Pidgins are very restricted structurally, and so have no real grammatical rules. Now, when the children of pidgin speakers are raised within a pidgin environment through the age at which they would normally acquire a native language, they spontaneously transform the pidgin of their parents into a grammatically complex 'real' language – and, of course, they do so without formal or informal instruction, because the pidgin is without structure. They just do it. No one tells them how. The result is a native language called a creole, a phenomenon much studied and reported on by Derek Bickerton. In recent decades a remarkable case of pidgin in deaf signers has been reported from Nicaragua, where sociopolitical circumstances dictated a swift change from a high degree of isolation of the deaf from one another to a situation in which they were brought together in deaf communities. The result was a sign protolanguage, a pidgin. Deaf children who experienced this pidgin through infancy promptly transformed it into a highly structured and expressive sign creole. The result was the creation of a new language within just a few years.

That deaf children who do not have access to conventional linguistic input, signed or spoken, develop gesture as a means of communication has been known for many years. In a recent cross-cultural study by Goldin-Meadow and Mylander, from the University of Chicago, deaf children raised in two different cultural and linguistic environments, one in the United States, the other in Taiwan, showed striking similarities in the language-like structure of their gestures.

All of these findings on deaf and normal-hearing children point to one conclusion: 'The evidence that language is an innate, species-specific, biological attribute that must possess a specialized neural infrastructure is so overwhelming, one might think that only those driven by some ideological agenda could fail to accept that' – to quote Bickerton's words. This does not mean that language in its entirety is a wholly specialized organ system that uses mechanisms only evolved in the service of language. As already said, we know that working memory is an essential part of language understanding and production, and we use working memory in almost all cognitive psychological functions. Nor does language necessarily operate using exclusively human-specific mechanisms. In a recent experiment human newborns and cotton-top tamarin monkeys (which are pretty average as primates go, and certainly not exalted apes) were identical in their ability to discriminate Dutch from Japanese sentences, but could not tell them apart if they were played backwards. This indicates that human sensitivity to certain properties of spoken language depends on features of the auditory system that are widespread among primates.

Another thing that is widespread among primates is the skilled use of forelimbs to a variety of ends. Everyone knows that when our ape ancestors evolved a bipedal stance, which they did around five million years ago, their hands were freed up, opening the way for the evolution of manual skills that, in modern humans, have reached a degree of sophistication no other animal can match. It is also well known that humans use gesture a great deal. Recent research by the American psychologist Jana Iverson and her colleagues has shown not only that gesture accompanies the speech of congenitally blind people, who have, of course, never seen gestures in their lives, but also that blind toddlers who are just beginning to acquire spoken language use gesture, and in a way that differs little from how sighted children use it. Could there, then, be a link between hands, hand gestures and language? There are strong advocates of this position, none more so than the late

William Stokoe, who pioneered the view that sign language is a true language with all the characteristics of spoken language. Here is a quotation from an e-mail from Stokoe:

Pan [the genus to which chimpanzees belong] and perhaps other apes know how to ask for food and grooming, using their upper limb, and facial and bodily expression, in ways others understand. It seems to me that *Homo* carried this much further, and **saw** that their gestures not only represented things (with the hand) and actions or changes (with its motion) but also the union of these two into a sentence (noun phrase + verb phrase). If that is what happened, sentence-like, noun-like and verb-like entities came into existence at one moment. An early species of *Homo* would have needed knowledge of space and time, and gestures pointing up and down, here and over there, or indicating in or out, go and come, and, metaphorically, now and before and after, would have been adaptations of great service . . . The facial and bodily expressions of emotions add adverbial and adjectival knowledge to the manual sentences. Thus a gestural and complete language could have developed, equipping early man with meanings and symbols of the language kind. After many tens of thousands of years, the developing mechanism of speech to stand for symbol-meaning pairs would have been the adaptation that ushered in the spoken mode of language.

This is a beguiling notion, and readers interested in pursuing it further should look at Frank Wilson's fine book listed at the end of the chapter.

In general terms, much is known about which regions of the brain are involved in language. About the details, though, virtually nothing is known. As to cognitive theories of the mechanisms that underpin language, there are, broadly speaking, two kinds. The one, owed largely to Chomsky, views language as the output of an innate organ of mind that develops pretty much like any other organ and functions according to a system of rules. Chomsky and his followers have been developing this line of thinking for over 40 years, its most recent formulation being known as the minimalist version of 'principles and parameters'. For Chomsky, the central problem is to discover the general properties of the rule system which, in specific language environments, finds expression in different languages. For example, some languages, such as English, are 'head-first', with the verb preceding the object, while others, such as Japanese, are 'head-last', with the object preceding the verb. In Chomsky's own words, from his most recent book on language:

We can think of the initial state of the faculty of language as a fixed network connected to a switch box; the network is constituted of the principles of language, while the switches are the options to be determined by experience. When the switches are set one way, we have Swahili; when they are set another way, we have Japanese. Each possible human language is identified as a particular setting of the switches – a setting of parameters.

Now, in order to work, the language organ must be embedded within the larger architecture of the mind/brain, and this requires that it satisfy what Chomsky calls 'legibility conditions'. These are of two kinds. One is a set of connections to the perceptual-articulatory apparatus, whereby sounds, or phonemes, can be expressed or decoded when heard (or, in the case of signs, when seen). The other is a set of connections whereby language gains access to the mind/brain's semantic processing apparatus. Thus there must be two interfaces, the one concerned with sound and the other with meaning, by which language can be 'read' and used by other mechanisms.

Chomsky's work has been hugely influential in shaping not only modern linguistic theory but modern cognitive science at large. It has led to the principal paradigm in modelling human cognition, based on the image of a complex of interacting subsystems, each with the specific function of processing symbols according to certain, in principle, explicitly specifiable rules. Crudely, the cognitive mind is a set of black boxes which process information and shuffle it between themselves, and the task of cognitive science is to identify those boxes and work out exactly what each does and what its role is in the overall system. Chomsky's version happens to be much more complicated than most (but then language itself is a complicated phenomenon) and has very strong nativist (innate) overtones. And it has been highly successful over these last four decades in seeing off general-process accounts of language, notably those proposed, quite separately, by Piaget and the American behaviourist B. F. Skinner. General-process accounts are identified by the assumption that language is not the output of a special organ of mind, merely a faculty determined by the same mechanisms as are responsible for our being able to associate one event with another, to imitate, to ride bicycles and to understand that when an object is obscured by another it continues to exist. Very few people today take earlier general-process accounts of language seriously. But a recent general-process account has proved more promising, and this provides the second set of theories currently being applied by those seeking to explain language mechanisms.

These theories were born out of attempts to explain cognition by simulating on computers how neural networks might operate, using models known as connectionist, or parallel-distributed processing, models. In essence these assume the existence of networks of initially randomly connected units, called nodes, the strength or weight of each connection, which determines the activity of a node, and the rules that determine changes in the strength of connection. Input to the network ripples through the nodes and produces an output, which is matched to a required output, the result of which is an adjustment to the weightings of the network. The procedure is iterative and results in the network being 'trained' to a desired state, in which the output of the network matches the requirement of the training. Learning thus occurs through the gradual change of connection weights throughout the entire network. Knowledge is the state of the entire network. Some success has been had in training such networks to recognize words, learn regular and irregular past tenses of verbs, and even apply regular past tense inflections to irregular verbs (*swimmed* instead of *swam*, for example), a form of rule-overgeneralization observed in normal language learning in children.

The assumptions of such dynamic models are wholly different from those of the symbol processing of mainstream cognitive science. There are no discrete symbols, no task-specific modules (or black boxes), and no explicit and specific rules applied to different subsystems, and processing is distributed through the whole network. Cognitive activity is a series of network transitions plotted mathematically. One of the advantages of connectionism is that it resonates well with what is known about the structure of real neural networks in real brains. It is easy to see how a node could be a neuron and the connections between nodes the synapses between nerve cells. There is also a close relationship between changes in weights or strengths in the connections between nodes and what neurophysiologists are telling us about how strength of synaptic activity changes with repeated activation of synapses. Finally, there is a conceptual link (evident in the very names) between connectionism and the associationism that dominated psychology for so long in the 20th century.

Needless to say, there is considerable tension between Chomskian linguists and connectionists. Nobody has yet trained a network to a state in which it has the quintessential feature of human language – generativity. It must be said at once, though, that the simulations use networks that do not begin to approach the complexity of real neural networks. And as so often is the case in scientific disputes of this kind, people tend to adopt extreme positions that work against the other side, and work also against any kind of rapprochement. For example, early in the development of connectionist modelling, the explicit assumption was made that the connections between nodes were initially random. But this is not an absolute requirement. Constraints, innate or otherwise, are easy to build into these models, and usually are. Connectionists are not necessarily radical empiricists and flag-bearers of the *tabula rasa*. And it does seem possible that different regions of the brain are characterized by parallel-distributed processing networks with different architectures of connectivity, different rules for setting node strengths, and different starting constraints, which would preserve the mainstream cognitive-science image of different boxes doing different things. This is not the place to attempt a reconciliation between these contrasting approaches to language mechanisms. It is likely that sometime in the future we shall have an understanding of language, and of other features of the mind/brain (because connectionist modelling has begun to embrace all aspects of cognition), that combines elements of both approaches.

I want to conclude this section with one final point. Despite our current inability to spell out with any certainty the details of language mechanisms, one thing is absolutely certain. Even if we do not know how language evolved and are unlikely ever to have this knowledge, language is, directly or indirectly, an adaptation or exaptation, a product of evolution. That being so, the capacity for language must reside *in part* – not entirely – in the genes (and please note the plural – it is probable that many genes are involved) of every one of us. These genes, bestowed upon us at conception, are, remember, the only point of contact each of us has with the evolutionary history of our species.

Theory of Mind

Every act of communication implies some degree of shared understanding and knowledge. In species that communicate without recourse to intelligence – those, for example, that produce and respond to innate alarm signals instinctively – what is shared is stored in the gene pool. In species that *do* draw on intelligence to communicate, for whom communication *is* an act of intelligence, what is shared resides in individual brains. Given that language, that unique form of human intelligence, allows us to consider the past and future and to communicate about things that are absent or even non-existent – given, in other words, the ability of language to dis-locate our thoughts from the immediate present – and given also that language is about communication between people, how can a speaker (or signer) and listener know that they both understand what is being said in the same way. Language alone cannot do this. There must be another feature of mind that operates to ensure approximate identity of meaning and reference if language is to be an effective form of communication, a feature of mind that has to have evolved in tandem with language, otherwise language would be little different from the signalling systems of other primates. This feature is what in recent years has come to be called Theory of Mind (ToM for short), mind-reading or mentalizing. It is the ability to attribute intentional mental states to others – in other words, to understand that others have a mind just like oneself, and that our minds work pretty much the same way in all of us.

There is nothing new about the notion that we are aware others have minds. Philosophically this is an idea that goes back to the ancient Greeks. It also appears in various guises in 19th- and 20th-century psychology, especially social psychology. But what is different about Theory of Mind in its present form is that it has become the focus of empirical study since the 1980s, especially in children, and the results have been so spectacular that it has moved centre stage in cognitive psychology, almost pushing language to one side. Briefly, this is the story. Up to about eight months of age, infants interact with the world, including other people, largely in terms of what Piaget called sensori-motor schemata. They are engaged in developing motor skills and the knowledge that comes with them – they move limbs, grasp and manipulate objects and use objects in conjunction with one other, as when repeatedly banging on a surface with a spoon. Then, what Tomasello calls the nine-month revolution occurs. Infants begin to engage with objects and other people in the form of a triad – self, other and object – where previously all interactions were dyadic, involving self and other or self and object. They are beginning to share attention with others, and this is a crucial first step in the development of mind-reading. Within a narrow cognitive developmental band between about nine and fifteen months of age, infants move from sharing attention, through following the gaze of others and checking its direction to ensure they are looking at the same thing, to declarative pointing. This, remember, is the period during which single- and then two-word utterances begin to appear.

At this point humans leave other apes standing still cognitively. There is simply no evidence at present that chimpanzees can attribute intentional mental states, such as knowing and wanting, to others in the way that we can. A clever series of experiments just published by Tomasello and colleagues used pairs of chimpanzees, one subordinate and the other dominant, in a specially constructed environment where they competed for food variously arrayed in space, sometimes behind opaque barriers such that some food items could be seen by only one of the animals. The results of the experiments showed without doubt that chimps know what other chimps can and cannot see and adaptively use this knowledge to their own ends. They can even tell which individuals know what. There is also clear evidence that many primates – and other mammals, such as carnivores – follow the gaze of others. But no apes apart from humans point to direct the attention of others. And there is no evidence that any non-human animal, chimp or otherwise, understands that others of its kind know things or want things in the same way that it itself knows or wants things. For example, humans can understand that others have beliefs different from their own and at odds with reality. Tomasello's chimpanzees, on the other hand, may only have known what their fellow subject knew. That is a significant difference, and, according to Tomasello, is possibly what prevents apes from truly imitating an action. Imitation requires the alignment of both a sequence of actions and an intended goal. If you cannot read the intentions of others, you are unlikely, except by chance, to reach that goal. It has been shown experimentally that children, observing an adult trying to do something without success and therefore not reaching the clearly signalled and intended goal, such as failing to separate an object into two pieces, will imitate the action and bring it to a successful conclusion. Apes cannot do this. This is probably also one of the reasons why they cannot acquire language either.

At around 18–24 months of age, children begin to enter into pretend play. If a child, in conjunction with one of its parents, zooms round the kitchen in a car that is actually a cornflakes box, then that joint pretence is evidence that the child has understanding of the pretended state of mind of the parent. At about the same age, children also begin to use mental-state terms, such as 'want', 'know' and 'need', with increasing frequency and attribute these mental states both to themselves and others. The defining point of Theory of Mind comes with the completion of its development somewhere around 40–50 months of age, when children arrive at the understanding that others can have false beliefs. Famously, this is demonstrated in a play situation in which a child observes two dolls behaving as if something in the world, such as the position of an object, has changed unknown to one of the dolls (Sally) but known to the other doll (Anne) and the child itself. When asked where Sally thinks the object is, children yet to reach their fourth birthday say that Sally knows what they themselves know. Repeat the test sometime after that birthday and children say that Sally believes the object is in the incorrect place. In other words, they have come to the understanding that Sally, and all others, can have false beliefs that may guide their behaviour. There is another way of putting this cognitive transition. Prior to age four, children think that what they know, feel and believe is what all other people know, feel and believe. After that age children can understand that others have minds of their own and that those minds have intentional states that might be different from their own. The next step is to use that insight to understand why others do what they do, to predict that others will do one thing rather than another, and, finally, to manipulate the intentional mental states of others in order to manipulate their behaviour.

Theory of Mind, like language, is widely held to be a unique, human-specific cognitive trait. This being so, it is assumed that, like language, all people, pathology aside, have Theory of Mind and that lacking it results in severe social pathology. In the 1980s the evidence from three British psychologists, Simon Baron-Cohen, Uta Frith and Alan Leslie, that autistic children generally fail the Sally–Anne test whereas age-matched normal controls and children with Down's syndrome pass it, was considered to be a triumphant affirmation of this position. Autism is a condition that only becomes easily recognizable after about the age of three and manifests itself mainly in impaired social, communicative and social-imaginative functioning. In fact autism, in common with most forms of mental abnormality, is not an absolute condition. Some autists have severe mental retardation, and some do not. Some show signs of Theory of Mind, and some show few or none at all. The specific causes of autism are not yet known but are generally assumed to produce malfunction of those parts of the brain/mind that enter into social intelligence, including Theory of Mind. Such malfunction can be more or less severe, hence the spectrum of severity of the disorder. Asperger's syndrome is a form of mild autism. Baron-Cohen and his colleagues have recently published a study of three high-functioning people with Asperger's, one a mathematician and winner of the Field Medal (the mathematicians' equivalent of a Nobel Prize, so you don't get much higher functioning than that), and two successful students (one in physics and the other in computer science). Three tests were administered to them. One drew on Theory of Mind, the subjects having to make judgements of the mental states of people in photographs. The second was a test of knowledge of folk or intuitive physics. The third was a very difficult test of reasoning ability called the Tower of Hanoi, which presents the problem of shifting discs around three pegs according to a simple set of rules such that one distribution of discs succeeds another. On the last two of these tests all three persons achieved the highest possible scores. But on the first test, which required social reasoning, all three were deficient, indicating that impairment in this area is independent of general intelligence and reasoning ability. It is a very encapsulated deficiency, indicating that Theory of Mind is a specific psychological mechanism with a specific neurological basis.

Brain-imaging studies support this view. Chris and Uta Frith report that thinking about the mental states of others involves specific brain regions. These, moreover, are regions associated with detecting the behaviour of others and analysing the goals of others, and activated by thinking involving mental states of the self. There is a striking association with areas 'relating to the representation of actions, the goals implicit in actions, and the intentions behind them' rather than the object-recognition systems of the brain associated with identifying faces and emotions. Piaget's 'in the beginning was the response' seems to apply to Theory of Mind too.

All such findings concerning Theory of Mind – that it is confined to humans, is localized within the brain and, especially, has a regular and rapid pattern of development – give support to the idea that it is an innate form of human intelligence in much the same way as Chomsky considers language to be an innate organ of mind. Certainly there is a strong nativist school of thought that considers this to be an important aspect of our understanding of Theory of Mind. Other explanations vary in their adherence to the notion of an innate, encapsulated form of intelligence. One, called the Theory Theory of Mind (no, the double 'Theory' is not a printing error) envisages the child as a kind of scientist. Evidence about the world is constantly pouring through sensory channels into the mind of the child. This evidence is organized into possible forms of coherent meaning by the child building theoretical constructs that explain the world and serve as a basis for predicting events. One form of evidence, and hence one set of theories, relates to what other people do and why they do it. In this way the child gradually arrives at an intentional Theory of Mind: 'People have minds just as do I,' thinks the child, 'and what those minds think, know and want are the basis of the actions taken by people.' A closely related position is simulation theory, which supposes children develop Theory of Mind by putting themselves in the position of others and simulating their possible internal operations, thus arriving at an understanding of what mental states drive others' behaviour.

A generally implicit, but sometimes quite explicit, assumption that all Theory of Mind theorists make is that Theory of Mind is a characteristic of all human minds. If this is correct, and if the mechanisms that drive it are the same in all people – and there is no reason to assume otherwise – then why, asked the American psychologist Angeline Lillard in the highly respected journal *Psychological Bulletin* in 1998, is it the case that the kinds of agency people regard as the causes of people's behaviour vary from one culture to another? For example, if someone I know passes me in the street and fails to acknowledge me, my assumption is either that I am guilty of having caused offence to that person and they are angry with me, or that they are so preoccupied with their thoughts that they did not see me. What I do not infer is that they are possessed by ancestral spirits and that it is these spirits that have caused the snub. Well, spirits as determiners of other people's behaviour do not have a place in my Theory of Mind, nor that of most people in the culture I live in. But ancestral spirits do form part of the causal attributions of the behaviours of people from other cultures. For instance, ethnographers have recorded that the Tallensi people of Africa believe that ancestral spirits live among them and cause people to behave in certain ways, and the Ifaluk of Micronesia also believe that spirits can and do enter the bodies of people and affect their behaviour. Anthropologists have reported many such instances of cultures that depart radically from the kind of Theory of Mind that dominates European and American culture. So, how to account for such differences of attribution? The answer to this important question is not simple and really involves several answers. First, there are some cross-cultural data that directly support the underlying assumption of universality. Study of the Baka people of Cameroon, who are non-urban hunter-gatherers has shown that four- to five-year-old Baka children pass the false-belief task and accurately predict the emotions of those holding the false belief. Similar studies in China and Japan, though of literate people living in industrialized cultures, also demonstrate the universality of the development of Theory of Mind. Well, these are promising findings, but we need much more work of this kind to be certain of the universality assumption, especially as regards the timing and sequence of the development of Theory of Mind.

The second answer is that individual cultures are not as uniform as Lillard implies. I do not believe that God is a factor in causing people to do certain things, but many of those who share my culture believe just that. So Theory of Mind does not operate to produce identical outcomes in people of (largely) the same culture. And the outcomes vary across time as well. European–American culture of a few centuries ago certainly did entertain the belief that people could be invaded by malign spirits and become witches (usually) or (sometimes) warlocks. It is trivially true that cultures are dynamic and changing. But, persists Lillard, if the mechanisms and processes are universal, why the differences in the content of Theory of Mind in people from different cultures? Lillard is admirable in giving many examples. Here are just a few. In Western cultures, folk psychology emphasizes that the individual is responsible for his or her own behaviour (though this is being diluted by the culture of blame), whereas in Japanese culture children come to believe early on that they are responsible for the behaviour of their peers. Hindu Indians tend to attribute behavioural causes to external situations, whereas in the West internal individual traits, such as resentment or ambition, tend to be emphasized as the cause. Even in the ranking of the importance of the senses there are differences. In Western societies vision is highly ranked, and much normal discourse involving understanding is couched in terms like 'I see that'. The Ongee of the South Pacific, by contrast, live their lives around smells and odours, to the extent that, as Lillard writes, 'even personal identity is wrapped up in the nose. Among the Ongee, one person greets another by asking, literally, "How is your nose?" ' Such differences must be explained.

Lillard, of course, is right in her demands and right in her assumption as to what the real answer is – an answer with which few would disagree, even if we do not know the details of how these differences come about. This, then, is that third answer. No one thinks that mental attribution, or mind-reading, is not universal. But Theory of Mind is not only essential for culture, it is also causally shaped by the culture into which we are enculturated. Western culture is dominated by the Theory of Mind that Western psychologists first identified when they postulated the existence of a Theory of Mind. That is, the emphasis is on individual needs and desires and how these are translated into action. This coincidence is not only unsurprising, but it would have been strange if it had been otherwise. But that does not preclude the same mechanisms operating in a culture that disapproves of publicly attributing intentions to others, such cultures being well documented, or which differs in its view of how people think. Ongee children are constantly exposed to parental references to odours. Perhaps instead of parents asking their children 'Do you see that?' when questioning understanding, they ask 'Do you smell that?' The result is that people in different cultures end up with qualitatively different Theories of Mind. In many cultures an important matter of custom and courtesy requires that people refrain from talking about the intentions of others and may not frame their Theories of Mind in terms of needs, wants and desires as explicitly as we Westerners do; and in other kinds of culture, people's thoughts are channelled so as to use different sensory input as metaphors. In short, if a mechanism is highly sensitive, from the very beginning of its operation, to the cultural milieu within which it operates, the outcome will be a form with content that mirrors that cultural milieu.

So the Theory of Mind mechanism is both shaped by culture and a crucial determiner of cultural continuity. It is essential for the deep communication of meaning both by language and imitation and it probably has a causal role in the growth of schemata and other higher-order knowledge structures. It also has, as will be argued in the final chapter, an essential role to play in the formation and maintenance of social constructions. But it is not immune to human creativity. People are, even if only exceptionally, able to break out of the Theory of Mind mould into which they are set by their early enculturation. We are able to discover models of the world, such as 'There is no God' or 'Everything must be done for the good of others', and adopt new Theories of Mind. If these creative inventions persist and spread, the culture changes, and with it the kinds of Theories of Mind possessed by its members. What never changes is our understanding that others do have minds, of whatever sort.

Social force

In the early 1950s, the American social psychologist Solomon Asch carried out a simple, and classic, experiment that was to give rise to hundreds of replications and extensions. He brought together a group of people and presented them with a vertical line on a card, and then asked them to say, one at a time, which of three lines on another card matched it in length. The match was easy to make, but the experimental arrangement was not what it seemed. Only one of the so-called subjects was truly an experimental subject, the other people being confederates of the experimenter, stooges planted to give erroneous answers on 12 of the 18 trials. (So, on six trials the confederates gave the correct answer, and on the remaining 12 they gave an incorrect answer, selecting a longer line six times and a shorter line six times.) The situation was also rigged so that the real subject was called to give their response after most of the wrong answers had been called. What Asch found, to his surprise, was that although about 25 per cent of the true subjects steadfastly stuck to their opinions and consistently gave the correct answer, around half of them, on at least half of the trials, gave the incorrect answer that corresponded with the staged and, to anyone with normal vision, obviously incorrect response of the majority. Overall, naive subjects conformed with the majority view on about a third of the staged trials. When Asch asked the subjects after the experiment why they had conformed, most said that they had known full well that what they had heard from the others had often been wrong, but that they had felt varying degrees of uncertainty and self-doubt and these had developed into feelings of anxiety and a fear of disapproval. They did not like going against the majority.

Subsequently, the experiment was run by Asch and others with many variations. One of these was to vary the number of confederates. The degree of conformity was originally found to rise as the numbers of confederates increased to three, but with little change thereafter; later studies reported conformity to increase steadily as the group size rose to eight. Other variations have included whether the group is made up of strangers or of people who knew one another, and the difficulty of the required judgement. By the middle 1990s it was possible to compare such experiments carried out in many different countries outside of North America (including about six in Europe, two in the Far East, and several in Africa, the Middle East, Brazil and Fiji). Conformity has always been found, but what is also clear is that there are cultural differences. This is not surprising. In societies that esteem individuality and independence of judgement, conformity occurs at a lower level than in cultures that value cooperation and agreement. It is also likely that the reasons for conforming differ across cultures. Not wanting to appear different might predominate in one setting, whereas tact and a desire not to embarrass peers might be the reason for conforming in another. So conformity might have different surface motives in different cultures, but a degree of conformity, on present evidence, does seem to be a universal of all human social groups.

Asch's work was an extension of experiments carried out by Muzafer Sherif in the 1930s. Sherif had used a visual illusion called the autokinetic effect to study how groups of people come to an agreement about an ambiguous event. The autokinetic effect is an illusion of movement which occurs when someone is asked to fixate their vision on a spot of stationary light in a darkened room. After a short time, we all experience apparent movement of the light – it seems to be moving, and people who are not told that it is an illusion truly believe that it is moving and can make judgements about the amount of movement. Sherif asked groups of people who did not know of the effect to estimate out loud the degree of movement they experienced. He found that groups would quickly home in on an agreed figure, which would eradicate the initial variation between individuals – people would report using the estimates of others as a frame of reference for their own judgements and thus a group norm would soon be established. Sherif's experiments might be thought of as demonstrating a form of constructive conformity, and indeed Sherif himself believed he was tapping into a fundamental feature of human social interaction. We each exert some kind of influence on others in a social setting.

Now, as already pointed out, Asch was surprised at his own findings because he had argued that Sherif's results were a product of the inherently ambiguous nature of the autokinetic effect and he had predicted that judgements made under conditions of clarity would not be influenced by the social force of others. Hence his use of simple stimuli and obvious differences in line length. Yet, as we have seen, even under clear-cut circumstances we may moderate our behaviour in response to that of others.

There is a twist to this tale. In the early 1960s a psychologist at Yale University, Stanley Milgram, considered that the Asch experiments were altogether too unrealistic and too tame as measures of social force. Much influenced by the horrors of the Holocaust, and especially by Hannah Arendt's claim that ordinary people living ordinary lives are able to commit unspeakable acts of evil against others provided that the acts are sanctioned by authority and meshed into the ordinariness of their lives – what Arendt referred to as 'the banality of evil' – Milgram set out to investigate just how far people will go when pushed by an appropriately oriented social force.

The situation Milgram set up was as follows: an unknowing subject, a volunteer from a very wide social spectrum of the male population of the city of New Haven, was brought into a psychology laboratory on the campus of the university believing that he was to take part in an experiment on the effects of punishment on learning and memory. Designated 'the teacher', this person was directed to administer electric shocks of increasing severity to 'the learner', semi-restrained in a chair and with an electrode attached to one arm, each time the latter made an error in a simple memory task. The shocks were administered through a device which clearly indicated their strength as ranging from slight, through strong, then intense, to 'Danger: Severe Shock' (at 450 volts). What the subjects did not know was that the learner, a mild, inoffensive-looking, avuncular figure, was an actor who never received any electric shock at all. But what the actor did was grunt, then shout, then scream and bang on the walls as the strength of the supposed shock was increased.

Prior to the first experimental run-through, Milgram asked a panel of people, which included some 30 psychiatrists, what they thought the result would be. There was near unanimity that none of the subjects would administer much more than a mild electric shock before defying the orders of 'the experimenter', the supposed authority figure whose power lay in his representing a respected university and the science that it pursued, and refusing to shock the learner any further. Well, they were wrong. Approximately two thirds of the subjects – ordinary people in the kind of situation Arendt had written about – completed the experiment by administering the maximum apparent punishment. Subsequent variations on the original experiment included the use of women subjects, with the same results, the learner professing a heart complaint, and the teacher having to force the hands of the screaming learner down onto a shock plate, thereby being physically responsible, or so the subject thought, for the punishment. The teachers did not do this without protesting, sometimes vehemently, at what they were being 'made' to do (although in fact there was nothing to stop them from walking out of the experiment at any time). And they did not do it without many of them being extremely upset. But do it more than half of them did.

One of the variants on the basic experimental arrangement was to have three teachers, two of whom were stooges, those old friends of experimental social psychologists, who, at prearranged points in the experiment, refused, first one and then the other, to proceed further and broke off and went and sat in another part of the room. In these circumstances total obedience fell to 10 per cent. Thus Milgram pitched conformity against obedience, and conformity proved, even under these extraordinary circumstances, to be a powerful force.

Milgram's results were shocking at the time – and still are, despite manifold examples of the banality of evil in places as far removed as Rwanda and the Balkans. Considered by many to be the most important psychological experiments in the history of the subject – and the transcripts of the verbal exchanges during the experiment are surely the most dramatically compelling ever yielded by an experiment – they starkly point to the power of social force. Milgram himself carefully distinguished between obedience, the primary object of his studies, and conformity of the kind reported by Sherif, Asch and others. In doing so he pointed out how differences in group structure give rise to different kinds of social force (obedience arises in the context of a hierarchical system of authority and power, whereas conformity occurs in the absence of such a structure), and compared the justification most of us give for obeying on the one hand (the infamous 'I was just carrying out orders', which shifts responsibility away from oneself) and conforming on the other (in which case we usually deny yielding to peers). Such differences are clear, but obedience and conformity are in essence merely variants of the same social force rooted in the group – in the collective of individuals who are the normal environment of the vast majority of human beings.

We know much less about the psychological mechanisms and neurological substrates of social force than we do about language and Theory of Mind. Recent studies have shown that damage to specific regions of the brain are associated with sociopathy, which, broadly speaking, removes people from the influence of both peers and authority. Sociopaths are laws unto themselves, which contrasts nicely with the way in which social force emanates from our responsiveness to the group and ties us to others – in just the opposite way to sociopaths, because for normal people the law is grounded in the group. One of the reasons for our lack of knowledge is that, in general, psychological science has focused far more on cognition than emotion, and social force seems to lie in the realm of emotion rather than cognition. If this is correct, then understanding emotion becomes part of a natural science of culture, because social force combines with Theory of Mind in guiding us in our understanding of what other members of our culture know and believe. We enter our culture at birth and become enculturated through a mix of formal tuition, which usually only begins after some years have elapsed, and intense if informal and less structured immersion in culture, which results in a 'soaking-up' of prevalent cultural knowledge, values and beliefs. We humans are social sponges, and part of the soaking-up process is driven by our propensity for obeying and conforming. Whether that propensity is a specifically evolved trait of our species is unknown. The great social scientist Herbert Simon considered this a real possibility. Well, perhaps so. In the next chapter we shall consider 'the group' rather more closely. Suffice it to say here that being social is characteristic of most primates, so cannot be a defining measure of what it is to be human. But a susceptibility to being influenced by others in terms of what we learn about the world and how we behave towards one another may well be one of the driving forces of our insistent sharing of knowledge and belief that does mark us out as different from all other species on the planet.

A single magical mechanism?

Could it be that in pointing to a set of, to some extent separate, psychological mechanisms I have missed the possibility that humans are uniquely creatures of culture because they have just a single mechanism, one potent driver of culture, which appeared just once in their evolutionary history, which underlies all or many of the mechanisms outlined above, and which transformed the species rapidly and irrevocably? Well, this is indeed possible, and there are several candidate mechanisms that could fill this role. Recursion and an understanding of discrete infinity, which we met earlier in this chapter in the section on language, might together be one such. Recursion is the ability of some 'thought entity' to invoke itself. It is what allowed Pinker so easily to break the record of the longest sentence ever produced – there is no such thing, because every sentence can become a phrase (albeit a long one) in another sentence. That is, every sentence can invoke itself. Recursion also occurs when 'I think that you want me to know that you need me to understand . . .' – this could go on forever, or at any rate to the limits of our working-memory capacity. Recursion operates even in 'simple' self-reflective thoughts or statements, such as 'Today I am happy.' It also enters our arithmetic understanding. There is a real possibility that the appearance, through evolution as an adaptation or otherwise, of the capacity to compute recursively is the key to some or all of the mechanisms that gave rise to culture.

Another possible candidate is some kind of supramodal or abstractive capacity that underlies our capacity for understanding higher-order knowledge concepts, as well as the linguistic properties of reference and meaning and our capacity for projecting knowledge of our own minds into those of others (Theory of Mind). This is an idea close to the French anthropologist Dan Sperber's notion that information coming into the mind/brain is processed in a first-order module specialized for a specific conceptual domain, the outputs of which are then processed in a second-order meta-representational module. (Meta-representations are representations of representations.) The first-order modules give rise to intuitive beliefs – for example, the belief that objects must be supported if they are not to fall – which are not specific to particular cultures. In contrast, the meta-representational module gives rise to what Sperber terms 'reflective beliefs' – for example, the belief that there is life elsewhere in the universe, or a belief in transcendentalist ethics or the theory of evolution – the constituents of which are not the products of a first-order module and which do differ across cultures.

Finally, there is the very fact of cognitive constraints. It might be a combination of one of the above together with constrained cognitive mechanisms – which have, in Chomsky's words, a 'rich innate endowment' of constraint that we share with others, allowing us to 'live in a rich and complex world of understanding shared with others similarly endowed' – that makes us the creatures of culture that we are. At present we have no way of knowing whether a series of separately evolving mechanisms synergistically drove each other on to the point where the capacity for sharing beliefs and knowledge emerged, or whether a single computational capacity is the key that unlocked this extraordinary feature of our species. But what we can be absolutely certain about is that culture, as an expression of human intelligence, is the product of a specific and knowable set of psychological and neurological mechanisms.

Suggested Readings

Baddeley, A. (1997) *Human Memory: Theory and Practice*. Hove, Psychology Press. (An excellent overview of current approaches to memory.)

Baron-Cohen, S., Tager-Flusber, H., and Cohen, D. J. (2000) *Understanding Other Minds*. Oxford, Oxford University Press. (Everything one might want to know about Theory of Mind.)

Calvin, W. H., and Bickerton, D. (2000) *Lingua ex Machina: Reconciling Darwin and Chomsky with the Human Brain*. Cambridge, Mass., MIT Press. (An exuberant argument about the origins of language.)

Chomsky, N. (2000) *New Horizons in the Study of Language and Mind*. Cambridge, Cambridge University Press. (A collection of essays from the originator of modern linguistics.)

Pinker, S. (1994) *The Language Instinct*. London, Allen Lane. (A wonderfully informative, if controversial, account of language.)

Rubin, D. C. (1995) *Memory in Oral Traditions: The Cognitive Psychology of Epic, Ballads, and Counting-out Rhymes*. Oxford, Oxford University Press. (An application of cognitive psychology to cultural traditions.)

Tomasello, M. (1999) *The Cultural Origins of Human Cognition*.

Cambridge, Mass., Harvard University Press. (A distinctive view of the links between culture and cognition.)

Wilson, F. R. (1998) *The Hand: How Its Use Shapes the Brain, Language and Human Culture*. New York, Pantheon Books. (A cultured and very readable account that links the hand to language and culture.)

Wilson, R. A., and Keil, F. C. (Editors) (1999) *The MIT Encyclopedia of the Cognitive Sciences*. Cambridge, Mass., MIT Press. (An exhaustive source of material in the cognitive sciences.)

6

Individuals, Groups and Culture

Social science is biology. It is complex biology. Some may like to think that it is special biology because it deals with what makes us human and different from all other animals. But biology it is, and so what a marriage between the social and biological sciences really boils down to is the incorporation of the key features of the social, that which makes us a uniquely different species, within the realm of biological science. Thus anyone taking on the task of pushing this marriage through had better build their efforts upon as solid a base of biological theory as possible. Incorporate the social into a biology that is wrong, or from which important parts have been left out, and the whole enterprise will be fragile at best; at worst, it will simply collapse. And it is not easy to get the biology right when the theory is evolution. No scientist worth her or his salt doubts that complex living systems are transformed in time and that the processes of variation, selection, and conservation and propagation of selected variants are at least partly responsible for that transformation. Nor is there any serious argument over whether life on Earth began about three-and-a-half billion years ago and that all contemporary living forms are connected by lineages that stretch back to those original life forms. Evolution is simply not in doubt. But there is real uncertainty about some key issues. The existence of adaptations is not questioned by anyone, but just how important are adaptations to the process of evolution? What exactly is the relationship between microevolution and macroevolution? How important are other factors, such as autogenous changes within genetic structures that are not subject to immediate selection? What really is the role of individual development in the evolution of species? Just how do we depict the conditions of evolvability under which living systems are sensitive to, and adjust in response to, changes in the world about them and yet remain stable in their transformed state? And just what are the units, the entities, the things that the theory picks out as significant parts of complex living systems and structures that we need to attend to, measure or describe in some less precise way? We are uncertain how to respond to all of these questions, and the assertion of some evolutionists that the answers, at any rate the really important ones, are now known does not sit easily with the history of evolutionary theory, which has changed in significant ways at least three times in the last century and a half. Who is to say it will not change again?

This is an especially pertinent thing to ask when one of the questions posed earlier in the previous paragraph is considered. That question, which asks what we should be focusing on as significant units or entities – what it is that the theory picks out as important – is about as fundamental as one can get. It asks what has consequence for the process of evolution, what we should be looking at and understanding, and what we should be measuring. Should we be concentrating on individual organisms, some part of organisms, such as organs, cells or genes, groups of organisms, or whole populations or species? The answer has changed over the decades, yet just what answer is given is obviously crucial when one is trying to relate the entities evolutionists think are important to the entities social scientists consider essential to their discipline. These are the connections that really do have to be made. This raises a set of issues that, to revert to our earlier analogy, are more biological threads that have to be woven into the tapestry before we can turn to the challenge of social constructions in the next chapter.

It is easier to understand why it is *these* biological threads that must be woven rather than others if the rather lengthy general argument developed in previous chapters is recapitulated. First, the marriage of the biological and social sciences must be founded on what we know about the human mind – that is where biology and social science meet. This is a commonplace and unoriginal claim. Second, the ceremony must take place within an ideational setting – we are, remember, following Goodenough's definition of culture as shared knowledge and belief. Third, and as a consequence of this ideational stance, culture is to be understood as a manifestation of human intelligence.

Fourth, we therefore have to understand something of the processes and mechanisms of that intelligence, which gives rise to the human capacity for creating and entering into culture. Now, the position presented in chapters 2 and 3 is that culture is an emergent property of certain features of intelligence (described in chapter 5); it is a 'supertrait' made up of some minimum number of necessary evolved components of our intelligence. In chapter 3 the strong assertion was made that the mechanisms which are causal to our capacity for culture had to have evolved first to some minimum level of competence before culture could appear. That has to have been the case. To claim otherwise is to put the cart of culture before the horses of its causal cognitive mechanisms. So, there has to have been a point in human history – and when this was we cannot know – when a non-cultural, or protocultural, human animal was transformed into a creature with some minimal capacity to share with others what it knew and did. We have called this earliest form of culture first culture. In essence, this is an emergentist view of culture in its original form. However, as pointed out in chapter 3, it is unlikely that culture as we know it now just popped out fully fledged at that unknowable time and has remained the same ever since.

We can never know with certainty what these founding mechanisms of first culture were. A reasonable guess is that at a minimum they combined a limited capacity for communication, perhaps initially simply in terms of copying the behaviour of others, with some understanding that others have goals. (Remember, we are considering the mechanisms essential for culture and not the panoply of mechanisms, such as the various forms of memory and communication, that operate when we function as cultural animals but which in and of themselves could not have given rise to culture.) Whatever these founding mechanisms, the consequences of sharing must have been advantageous to the members of the group doing the sharing. This, of course, is an adaptationist claim that cannot ever be verified. But in a relatively long-lived species, such as these ancestral humans would have been, with offspring born in a state of helplessness and therefore requiring a prolonged period of care as physical and psychological abilities develop over months and years in an uncertain and complex world, being able to learn through the experience of others, being able to instruct or teach and hence to pool skills and knowledge, had to have improved the chances of surviving and thriving. Early humans, though long-lived relative to most other animals, were not on average as long-lived as we are now, so the interval between conception and reproductive competence approached half a normal lifespan. For so active and socially interactive a species, that means each individual had to deal with a lot of change and short-term stabilities (to use the language of chapter 2), the mastery of which would have been essential to survival. Well, broadly speaking, there are two ways of gaining such mastery. You can do it entirely through your own learning, or you can do at least some of it by learning from others. This does not mean learning *because* of others, i.e. indirect social learning such as results from having one's attention drawn to specific features of the world, but learning directly *from* others – and learning from others what they in turn have learned from yet others. One can argue about the boundaries of the conditions that determine the point at which learning from others becomes advantageous over learning entirely from one's own experience, and one can model these as Boyd and Richerson, among others, have done. The results have been consistent. In a world of very rapidly changing circumstances learning from others is not a safe bet because one may acquire information that is out of date, perhaps partially wrong or even completely incorrect. However, in a relatively stable world, especially one in which parents and other adults are the principal agents of stability, learning from others, especially from adults, becomes an adaptive strategy.

Thus no one denies that under certain circumstances culture – learning what others have learned – must have been a good thing in the sense of increasing the biological fitness of those making up the sharing group. In which case first culture, with its constituent processes and mechanisms, would itself have become the target of selection. Those founding psychological processes and mechanisms that were causal in first human culture would have been further tuned and sharpened; additional mechanisms that reinforced the capacity for culture – perhaps language as a device for imparting massive quantities of precise information, and responsiveness to social force – would have been selected for. As a result, culture would have been transformed into an even more powerful force of human thought and action. Whatever that first culture was, it was certainly different from what human culture had become by, say, ten to twelve thousand years ago, when the agricultural revolution began. By then, humans had a linguistic competence very close to any modern language; we lived as easily within the intentional minds of others as we did within our own; and we were effortless constructors and communicators of higher-order knowledge structures. These changes from first culture to the beginnings of modern culture, like all transformations of complex organic features, had to have been driven by evolution.

But here is a problem – perhaps *the* problem. Culture, by definition, involves more than one person. Culture is a property of a group, and, as mentioned in an earlier chapter, moving from selection acting on individuals to selection acting on a group, or on properties that only exist by virtue of individuals making up a group, has been a hotly contested issue among evolutionary biologists over the last 30 to 40 years. We need to consider if group-level properties raise difficulties for evolutionary theory. First, though, we need briefly to review how biologists have thought about groups over these last few decades without invoking the complications of group-level selection.

The behavioural ecology of group living

In the early 1970s William Hamilton wrote a charming paper concerning the 'geometry of the selfish herd'. It told the story of a pond of frogs that was home to a frog-eating snake. The snake was an obsessive timekeeper as far as eating went. Every day at noon it ate one frog. Now, because catching frogs was much easier for the snake when the frogs were in the water, and because none of the frogs thought to make life simpler for the snake, at two minutes before noon each day (yes, the frogs also had wristwatches and were only too well aware of the snake's obsession with time) the frogs scrambled out of the pond and stood along its edge. (They didn't flee into the surrounding woods because there were monsters there who did far worse things to frogs than just eat them.) Following their scramble out of the water, the frogs found themselves distributed at random round the margin, and knew that when the snake rose to the surface the frog most likely to be seized and eaten was that with the largest domain of danger. The domain of danger is the distance from either side of each frog to the point halfway between it and its two nearest neighbours. Frogs furthest from their nearest neighbours had the largest domains of danger, while those closest had the smallest. So what each frog did from the moment it jumped out of the water until the snake rose and grabbed its midday meal was make strenuous efforts to reduce its domain of danger as much as possible. It did this by leaping over the neighbour whose own other neighbour was closest to it and tucking in between them. This dramatically reduced the leaper's domain of danger and hence the likelihood of its imminent demise. But the neighbour was no fool. Realizing its own domain of danger had been increased by its neighbour's selfish act, it too leapt, following the same reasoning, over one of its neighbours – and all the frogs were doing this at the same time. At the end of two minutes, what the snake saw as it rose to the surface was a heaving heap of frogs – or perhaps several heaps, depending on the initial distribution round the pond's edge. It would grab a frog and sink back into the depths, leaving the remaining frogs to breathe a sigh of relief that the victim had been someone else, disentangle themselves and return to the water, safe for another day. Hamilton provided a mathematical model of this grouping behaviour based on the simple instruction obeyed by each frog: reduce your domain of danger by reducing the distance between yourself and your neighbours.

The story illustrates several things. The first is that Hamilton was adopting the familiar stance of asking about the adaptive advantage accruing to specific behaviours of an animal – why is it doing what it is doing? One of the most common things animals do is live in social groups rather than as solitaries. So, what exactly are the advantages of living in groups? The second is that Hamilton was supplying a specific answer to this question. One of the reasons some animals live in groups is as a defence against predation, by, as it were, diluting their degree of danger by hiding behind their neighbours so it is they who are at risk rather than themselves. The third is that the reasons given for group living are almost always in terms of the benefits it has for the individual. The group is not seen as *an entity*, as a functionally organized whole with characteristics that make it capable of being something for which circumstances could be 'for the good of', in the same way as biologists routinely think of individual organisms being something for which circumstances could be 'for the good of'. It is the last point that needs emphasizing, but before expanding on it, it should be said that dilution of danger from predators is not the only answer as to why animals live in groups.

Good empirical studies have shown that the distance from a flock of pigeons at which a predator is detected increases with flock size. Improved vigilance, in other words, is another reason for group living. Lots of eyes are better than a single pair, and with increased vigilance individuals in the group have more time to deal with other needs, such as foraging for food. Group defence is another, obvious, benefit of group life. One baboon is no match for a leopard, but several may drive a leopard away. Mobbing behaviour in many species of small bird is another example of group defence.

There are yet other advantages to group living. Some predators, such as hyenas and Cape hunting dogs, are highly social animals, and their hunting is remarkably efficient precisely because they hunt in coordinated groups. Animals like birds and bees forage efficiently because group life allows for sharing of information and hence more effective coverage of space. Group living also facilitates exchange of sex cells in sexually reproducing animals and is a natural consequence of parental care of offspring.

Group living does have its costs, though. Echoing the arguments during the Second World War among naval strategists who were divided over the merits and disadvantages of merchant ships crossing the north Atlantic in convoys, and who offered alternative models in support of one or other position, behavioural ecologists have weighed the costs of the conspicuousness factor of group living against the advantages of group defence. Their models can become very complex indeed, depending on how many features of the groups, and of the individuals of which they are composed, are taken into account. Sensory acuity, speed of response, and strategy, in respect of both prey and predator, are just some of these, and if too many are plugged in the equations become almost intractable.

Other costs of group living include the ready transmission of infectious diseases, cannibalism and the intrinsic costs of sharing resources or having them stolen. One of the consequences of the last set of factors is that groups may develop complex structures consisting of subgroups, factions and cliques based on age, sex or kinship. Primates, especially monkeys and apes, present excellent examples of how the trade-off of costs and benefits in different ecological circumstances results in complex social-group structures. These structures drive demands on intelligence. The histories of who has done what to whom may become the basis for predicting who will do what to whom in the future. Thus life in social groups may generate selection pressures for the evolution of better or more specialized cognitive skills. This is often referred to as the social function of intellect hypothesis, which was developed independently by American primatologist Alison Jolly in the 1960s and N. K. Humphrey, an English biologist, in the 1970s.

Thus it was that understanding of groups within the evolutionary framework of costs and benefits, and how these vary with ecological circumstances, advanced apace over the last 40 or so years. The pluses and minuses of group life were considered overwhelmingly in terms of the individuals making up the groups. It should be added that group living, of course, long predated the appearance of culture. With the odd exception, most species of monkey and ape are social animals. Both our nearest living relatives, the chimpanzee and gorilla, are social. It is most likely, therefore, that the Miocene ape ancestral to the chimpanzee and human lineages was a social animal too. While once again we cannot be absolutely certain, it is reasonable to assume that for over five million years human evolution took place within the constant context of group life and its accompanying costs and benefits.

The units and levels of selection

What the units of evolution and selection are, and how they relate to one another, is an issue that has already been raised in chapter 4, but it is of such importance to the marrying of the social and biological sciences that no apology is made for raising it again. From that chapter it will be remembered that Darwin, being a fine scientist, had, in the years of research that preceded publication of *The Origin of Species*, searched through what was then known about the natural world to see if he could find phenomena that would test to the limit, perhaps even discredit, his theory of evolution. One of the phenomena he concerned himself with was sterile castes of social insects, because sterile insects cannot have offspring, and hence might have been thought to have posed him a problem, for his arguments were couched in terms of the individual. He wrote, for example, that 'there must in every case be a struggle for existence, either one individual with another of the same species, or with individuals of different species'. The entity that gains from such struggle is the individual, whose benefits are cashed out as offspring. It is the individual that profits. Unable to play this game, sterile insects, bristling with adaptations, must surely present a problem. But 'if such insects had been social,' wrote Darwin, 'and it had been profitable to the community that a number should have been annually born capable of work, but incapable of procreation, I can see no especial difficulty in this having been effected through natural selection.'

Darwin, probably unwittingly, had shifted the focus of his theory from the individual to the 'community' or group. The adaptations of the individual sterile insect are not, in procreational terms, good for it, but they are good for the group of which it is a member. At the time, and indeed for something approaching a hundred years, this switch from what is now called an individual-selectionist position to one of group selection raised neither eyebrows nor comment. In fact, with increasing acceptance of evolutionary theory in the first half of the 20th century it was not uncommon for group selection, in the form of it being the species that was the beneficiary of individual adaptations, to be proposed as causal evolutionary explanation. Indeed, often enough it was not just the species that was deemed the beneficiary but communities of species and even the entire biosphere. Explanation was often as focused on the upper levels of an assumed and implicit hierarchy as on the lower levels. Historically, the tendency towards group-selection arguments increased because of the influence of systems theory, which was widely discussed in the 1940s and 1950s. Systems theory, as the name implies, was concerned with understanding complexly organized entities, such as organisms, groups of organisms like social insects, and human organizations such as companies or local communities, with an emphasis on understanding such entities as integrated systems – as wholes rather than separately functioning parts. Hierarchies were widely held to be central to complex systems, and as we saw in chapter 3 with Simon's parable of the watchmakers, structural hierarchies in particular were thought to be important in biological systems. Cells, organs, individual organisms and groups of individual organisms made up one commonly invoked structural hierarchy, and with evolutionary theory as a constant conceptual backdrop to systems theory, the notion of adaptations was worked in at all levels of the hierarchy, especially at the highest level, that of groups (and groups of groups), without regard to how adaptations at one level might be in conflict with adaptations at other levels.

The clearest instance of such thinking came in a book, published in 1962, by the Scottish biologist V. C. Wynne-Edwards. Working his way through many examples of social behaviour in animals with admirable clarity, Wynne-Edwards consistently postulated a group-selectionist argument. Species of bird that form large flocks in the winter, for example, would adjust their subsequent breeding behaviour in accordance with their numbers and the availability of resources, the outcome being good for the group. No cognitive calculations by the birds was posited. The outcome was caused by the physiological effects of temperature, crowding and levels of food resources; combinations of such factors, such as low temperatures and high numbers, would reduce, it was argued, the overall fertility of the birds, or, if the conditions were reversed, would increase it. No decisions, conscious or unconscious, mediate the effect, which is the result of the responses of individual birds to stress. The net outcome is good for the group.

Others had been less restrained than Wynne-Edwards. Functional organization into highly integrated adaptive groups and populations was seen by some as a law of nature, with explanations verging on the mystical. 'Naive group selectionism' is how this movement is now frequently described. It should be noted that such thinking was not directed with much frequency towards human behaviour. Humans were not the point. It was the world of nature at large that was the target. But it was the lucidity of Wynne-Edwards' arguments, couched in terms of plausible physiological mechanisms, that brought group-selectionist thinking under a critical scrutiny that had been largely absent since Darwin's original explanation of sterile insect castes.

As touched on in chapter 4, the weakness of the group-selection position is that what is good for the group might not be good for the individuals making up the group. In other words, adaptations at one level of the structural hierarchy might be at odds with adaptations at other levels. Biologists like G. C. Williams in the United States and John Maynard Smith in England argued not that group selection is impossible (indeed, in his classic 1966 book, Williams concedes the possibility that at least one experiment by Lewontin with mice provides evidence of the operation of group selection) but that it is unlikely. In the example of overwintering bird flocks, mutant individuals are bound to arise in time – and remember that evolution has a great deal of time to play with – and one such mutant might lack the necessary physiological mechanisms for linking stress responses to fertility. Such a mutant will breed without regard to weather and resources, with the inevitable result that it will on average produce more offspring each year than birds breeding under the restraints imposed by winter conditions. Assuming such unrestrained breeding is an inherited trait, in time such mutants will come to dominate the flocks, and what was good for the group will have been displaced by behaviour that is for the good of the individual. So, while group selection is a possibility in theory, for the group to be a unit of selection requires that groups compete with other groups, just as individual selection involves competition between individuals, with some groups outcompeting others. It also requires, it was argued, negligible migration between groups and the failure of mutants arising within groups whose behaviour would undermine the group-level adaptation. Such a combination of features would occur so rarely that group selection as a general phenomenon of any biological importance was dismissed.

The seminal papers of Hamilton, which also appeared in the 1960s, further undermined the group-selectionist position. In very simple terms, what Hamilton wrote, in formal mathematical analyses, was an account of gene transmission across generations via two possible routes. One is individual fitness and individual selection, whereby organisms ensure the transmission of genes to their own offspring. The other is inclusive fitness, whereby an organism helps to propagate the genes of genetic relatives (hence the term 'kin selection'), some proportion of whose genes are identical to those of the helper organism.

Directly or indirectly, it is the genes that get propagated, and when this model is applied to Darwin's problem of the sterile insect castes, the adaptations of the sterile insects make sense within the framework of the adaptations being for the good of the genes of the sterile animals rather than being for the good of the individuals bearing those genes or their being 'profitable to the community'. To that curious metaphysical question raised in chapter 4, 'What is life all about?' (in the scientific sense of course), the answer on offer is that life is about genes. Hamilton's ideas did not deny the possibility of group selection. What they did do is focus attention on the gene as the unit of selection.

Inclusive fitness and kin selection took the community of scientists studying animal behaviour by storm. Within a decade the empirical and theoretical landscape of animal behaviour had changed. From parent–offspring interactions to fatal fighting, many forms of social relationship and social behaviour were being explained in terms of the genetic relatedness of individuals. In the language of the historian of science Thomas Kuhn, a paradigm shift had occurred, with the conceptual change driving new approaches to studying animal behaviour. Running alongside these events was the introduction of game theory into evolutionary biology. Game theory, which is central to much of contemporary economics, is a form of analysis that considers agents interacting with each other, the interaction being a game. Agents, which are thought of as rational decision-makers striving to maximize their own utilities, i.e. trying to do what is best for themselves, have strategies and make decisions about how they will interact with other agents. Agents are not necessarily individuals, they could be companies, military units, or even whole countries, where decision-making powers are vested in governments. However, the obvious thrust of economists using game theory is to consider the players as individuals or other economic units, such as competing supermarkets or banks. The point of game theory is to model the outcomes that arise when agents with different strategies play games of different kinds against each other, and hence to understand the optimal solution for agents in terms of what strategies they should be using in the context of the strategies of other agents.

When Maynard Smith and others transported game theory into the realm of evolutionary biology, specifically animal behaviour, the general assumption that was made was that the agents were individual animals interacting with other animals, and that the strategies were behaviours that had been put in place by evolution to maximize individual or inclusive fitness. Unlike economists, who have no problem talking about the pricing strategy of one supermarket chain as it competes for market share with another supermarket chain – in other words, with agents that comprise aggregates of individuals who are able to arrive at collective, consensual decisions – biologists are denied the freedom of thinking of the agent either as a single organism or as a collective of organisms. You can't think about wildebeest or digger wasps in terms of collective decision-making. That, after all, is the whole point of this book – it is only human beings that can make collective decisions. While it is not an absolute that game theory when applied to animals must be at the individual level, this is certainly the obvious starting point. So evolutionary game theory, Hamilton's insights and the arguments against the likelihood of group selection being a stable state, all combined to emphasize further the role of the individual and the individual's genes in the evolutionary process.

There is a further point to make, which concerns what one might call the principle of the importance of enduring. G. C. Williams' 1966 book, *Adaptation and Natural Selection*, which came to have enormous influence in the years following its publication, was an elegant and searching analysis pitched against group selection, among other things. What Williams wanted to do was banish sloppy thinking of all kinds from evolutionary biology and introduce a sharpened reasoning into the uneasy concept of adaptation – to make it quite clear what can be an adaptation and what cannot (the difference between a 'fleet herd of deer' and a 'herd of fleet deer' is an example of the kind of distinction he was trying to draw), and what the source of these adaptations is. He was thinking in the 'conventional' mode, i.e. in terms of evolution occurring by natural selection sifting and sorting between variable phenotypes with varying fitnesses over long periods of time and many, many generations. The fittest organisms – those that are best adapted to the environments in which they live – transmit more of their genes to subsequent generations because they have more offspring. For Williams, as for all conventional evolutionists (by which I mean those whose concern is with biological evolution pure and simple, rather than universal Darwinism with evolution of several kinds occurring both within and between organisms), an adaptation is a trait that spreads throughout a population, and is then maintained as a species characteristic for thousands, hundreds of thousands or even millions of years. An adaptation is not, for conventional evolutionary theory, a fleeting, fly-by-night characteristic. It must *endure*. Adaptations are not transient characteristics of just a few members of a population. They are widespread and stable traits. Pathology apart, all humans have language; all deer are fleet of foot; all predators stalk the fleet-of-foot deer in ways that bring them as close as possible to their prey before they are detected – and in each case, these are traits that have endured for hundreds of thousands of years at a minimum. Such sustained traits that enhance individual fitness, i.e. such adaptations, are an important means – probably not the only means but one of them – by which evolution occurs. And once an adaptive trait has been selected for and fixed within a population, it will, pathology aside, remain a trait in every future member of that population for as long as it does its job of contributing to the fitness of each individual – and perhaps even beyond then, as in the case of the human appendix. Now only genes have the requisite staying power in terms of endurance of structure to maintain such enduring adaptations. Only genes are constant across geological time (mutations are a complication but they do not alter Williams' main point). The genotype, which is the total genetic make up of an organism, lasts only from conception to death – a single lifetime – and is drastically altered by the reducing division – meiosis – that occurs during the formation of sex cells. (Asexually reproducing organisms, which form as clones of their parents, are an exception, but again, they do not detract in any way from the thrust of Williams' argument as applied to sexually reproducing organisms, which is what most chordates are.) The phenotype, the flesh-and-blood organism, obviously has no greater endurance. Neither, argued Williams, does a group of organisms. Organisms come and go, through death or migration, so groups are as transient as individuals. Only genes endure long enough to provide the consistency of something – call it information if you will – for maintaining adaptations through geological time. Socrates' genes, remarked Williams, may yet be with us, but not his genotype, nor his phenotype, nor the groups of which he was a member in ancient Greece. For this reason, Williams considered genes to be the primary units of selection, perhaps the only units of selection if one excludes universal Darwinism. They have 'a high degree of permanence and a low rate of endogenous change, relative to the degree of bias (differences in selection coefficients). Permanence implies reproduction with a potential geometric increase.'

It is because genes have this quality of enduring that Williams said, later in his book, that 'Ontogeny is often intuitively regarded as having one terminal goal, the adult-stage phenotype, but the real goal of development is the same as that of all other adaptations, the continuance of the dependent germ plasm.' As we saw in chapter 4, it was Dawkins, with his penchant for universal Darwinism, who extended this way of thinking by developing the notion of replicators – entities capable of copying themselves, which is how they endure and drive evolution and come to be what all adaptations are for, irrespective of what kind of evolution one is talking about, biological, cultural or any other.

This hardening focus on genes, because they are replicators, left something of a hiatus in evolutionary theory for the rest of the structural hierarchy that extends from cells (and cellular organelles), through tissues and organ systems, to organisms and various groupings of organisms. Are these all unimportant? If so, then replicator theory really would be a radical departure from orthodox NeoDarwinism, the standard view having been that evolution occurs in a two-phase process: selection acts upon the phenotype, variable forms of which are a consequence of development from a genotype, and the fitness success of the phenotype results in the genes of that phenotype's genotype being retained within the population gene pool. In other words, the phenotype is a kind of test bed of the suite of genes in the genotype, and there is an indissoluble link between the two. Genes may be wondrous things and they do indeed endure, but can they bear the entire weight of causal explanation of the processes of evolution?

They cannot, of course, but that didn't stop Williams from considering them, on account of their endurance, to be *the* fundamental unit for anyone thinking about evolution. That phrase, *thinking about evolution*, is important. Dawkins' *The Selfish Gene*, first published in 1976, was felt by some biologists to be a too extreme and stark expression of Williams' stance. In response to those critics, Dawkins wrote a more technical book called *The Extended Phenotype*, published in 1982, in which he presented a lengthy defence of his ideas – and extended them as well. One of his points of defence was in response to those who felt that once again evolutionists were downgrading the importance of development. He acknowledged that genes are just one of the many causes that enter into development, and he readily conceded the role of complex within-cell and between-cell interactions. When one is thinking and talking about development 'it is appropriate to emphasize non-genetic as well as genetic factors,' wrote Dawkins, but when one is talking about evolution and the units of selection, the emphasis must be on genes as the units of selection because they, and only they, are replicators in biological evolution. Only they retain their integrity, only they endure, in a way that no other element in the structural hierarchy does. So, when it's evolution that's under consideration, the central cause is one thing, and when it's development, other factors have to be considered: '. . . legitimate stress [must be placed] on considerations which are important for two different major fields of biology, the study of development and the study of natural selection'. Well, perhaps he had a point. Perhaps many had been overzealous in extending the explanatory scope of evolutionary theory, which cannot provide understanding of everything in biology, and certainly not development. But it still doesn't quell doubts about an artificial cleavage between genes on the one hand and what selection acts upon on the other, because the causal linkage between the two is absolute. This causal linkage is why some people believe development cannot be dissociated from evolution; and why some find overblown and arbitrary the claim that genes (or other replicators), above all other entities, are 'what evolution is about'.

Now Williams always held that selection can and does operate at all levels of the hierarchy, and that adaptations can and do form at all levels – even the group – but, for reasons outlined a few pages back, he thought group-level adaptations unlikely. There are many *levels* of selection, but only one *unit* of selection – the gene. Dawkins' conception was a little different, though not in essence. For him replicators often replicate better when enclosed within a surrounding structure, a cell. Simple single-celled life forms, technically called procaryotic forms, that lack a membrane-bound nucleus for containing the replicators, like blue-green algae and bacteria, first appeared on Earth somewhere between three and three-and-a-half billion years ago. Some replicators found replicative advantage if they were contained within a membrane-bound nucleus inside the cell, so by about 1.2 billion years ago such eucaryotic single-celled organisms had evolved. It took a further five to six hundred million years for replicators to 'find' that replication was facilitated by encasement in cells that linked together to form multicellular life forms. Whether single-celled, with or without a nucleus, or multicellular, all such entities were described by Dawkins as vehicles, which was an appropriate term given his conception of replicators riding around in rather passive entities. For all the reasons already given, Dawkins did not accord vehicles much causal force in evolution, and in *The Extended Phenotype* he went even further when he considered replicating genes as having causal force that goes beyond the individual bodies in which they reside. For example, a beaver fells trees, which has the effect of creating a dam. As far as is known, this is not a learned behaviour but one which is innate and hence part-genetically caused. It is also a behaviour that is advantageous to the beaver, which can now forage along an increased length of shoreline without having to make long and possibly hazardous overland forays for fresh resources. The dam is an extension of the beaver's phenotype and is as much a product of the beaver's genes, its replicators, as are its gnawing teeth or fur. The same can be said for many other animal artefacts, such as birds' nests and termite mounds. Indeed, the same can be said for the behaviour of any animal caused by the innate, even if part-learned, behaviour of another animal. Thus, when a bee, or a group of bees, forages in a certain locality because of the dance of another bee in the hive – which is how bees communicate the whereabouts of food sources to one another, the general form of the dance being innate and the specific form relating to position in space – that foraging behaviour is an extension of the dancer's phenotype. The analysis is consistent, and by extending the phenotype in this way beyond the organism, Dawkins seems to make an even more potent case for the causal force of genes while reducing the causal force of the vehicles.

The position sketched over these last few pages has attracted an enormous amount of attention, and polarized opinion, within biology. I know of no careful survey which reveals whether the genes-as-replicators view, with its implication that replicators are the dominant force in evolution, has become the majority view amongst biologists. It has certainly had a powerful impact in certain quarters, especially among those studying animal behaviour and evolutionary psychology. And it has also raised a storm of criticism. Some of this has taken the form of unhelpful caricature, which has fed on the fuel of reification, the creation of something that does not actually exist, in which some gene-centred theorists have indulged. One continues to read accounts in the popular press, for instance, of men finding women with a particular waste-to-hip ratio sexually attractive because, for a whole host of possible reasons, such women are likely to be more fertile, which is taken to illustrate how men's sexual preferences are pro-grammed by their genes, whose sole need and purpose is to propagate themselves. Genes, of course, have neither needs nor purposes. And before me is an article put out on the Internet by the Associated Press entitled 'Is there a music gene?' This cites the ramblings of a professor of music concerning the evolution of music in human affairs, who seems to have little understanding of biology and the unlikelihood that a single gene governs human musicality, but just enough sense to cover himself by noting 'of course it's utter speculation'. So then, why speculate and speculate so badly? And why do the mass media insist on propagating such silly ideas? This is fertile ground for a sociology of lay knowledge and not a question that can be pursued here. But as indicated in chapter 4 in the discussion of memes as replicators, the attribution of such properties as intentional agency and the overexten-sion of the causal force of replicators are both unnecessary and incor-rect. Replicator theory is powerful enough and need have no recourse to this fanciful baggage. But powerful as replicators may be, they are not the whole story. In answer to the question of what entities evolutionary theory should be focusing on, replicators, like genes, should certainly have our attention – but not all of it.

Vehicles, interactors and the revival of group selection

Simply put, replicators in the form of genes do not have a monopoly on causal force. The vehicle, which is usually the phenotype, extended or otherwise, is not passive. It is the focus of multiple interactions, both within itself (intracellular, intercellular and inter-organ) as a coherent functional unit, and between itself and the world of which it is a part, both social and otherwise. If it is a vehicle of a species that has evolved intelligence, then, as shown in chapter 2, it is a vehicle that gains information and makes decisions on the basis of that information about where it lives, what it eats and with whom it consorts and mates, decisions that cannot be reduced to genes – intelligence, remember, is a proxy information gainer and utilizer. Intelligence might, and usually does, act 'on behalf' of genes, and as an innate capacity (or set of capacities) must be part-caused by genes, but that does not mean that either intelligence in its many forms, or its consequences, can be causally reduced to genes. Dynamic, highly interactive, and when intelligent the source of the causes of its own behaviour, the vehicle is something much more interesting than the word implies. Hence the more appropriate choice of *operator* by Waddington and *interactor* by Hull, with connotations of dynamism, control and goal-directedness. Waddington's *operator* might be preferred to Hull's *interactor*, but because Waddington's ideas in this regard were largely ignored and Hull's *interactor* now has wide currency in biology, it is *interactor* that will be used here.

But just what is an interactor? Again, this raises the question of what evolutionary theory should be focusing on. Are the interactors always whole, single organisms and nothing else? Or is there room for thinking about some of the things that make up organisms, such as organs, as interactors, or, most contentiously, for considering groups of whole organisms as interactors? Since it is the group that is inseparable from culture, it is this on which we will concentrate here. Put simply, can a group of organisms ever have characteristics such that it equates with a single organism – or a superorganism, if one prefers? There are two ways of answering this question. The answers are related but sufficiently different that for the sake of clarity they will both be given in full. One concerns whether groups can have characteristics that can be selected for, and hence which can evolve. The other is a matter of part–whole relationships.

Let's deal with the selection and evolution issue first. As noted in chapter 4, and earlier in this chapter, the notion of group selection was very nearly killed off in the 1960s. A small group of mostly American biologists, David Sloan Wilson and Michael Wade among them, had the courage and insight to persist with the notion, and through their efforts there is now increasing acceptance that group selection can be, and has been, a powerful force in the evolution of life on Earth. Because this is necessarily a brief account of what are complex and controversial matters, readers wanting a whole-book treatment should refer to the volume by Elliott Sober, a philosopher of biology, and D. S. Wilson himself, a work to which this chapter section is much indebted.

Perhaps the clearest example to use is the evolution of virulence, which Lewontin, in his classic 1970 paper, pointed to as a test of group selection. The reasoning goes like this. If the survival and reproduction of the individual is the principal means by which genes are replicated and propagated, then organisms like bacteria and viruses should evolve to maximum virulence. This is so because the organisms that infect a host animal or plant themselves have genetic variability on which natural selection acts. The infectious agents compete with each other within the host, and the most successful reproduce at higher rates with consequent deleterious effects on the host. The extreme result is the death of the host, and unless this is an essential stage in the life-cycle of the disease agents, it means the death of the agents too, which is not good evolutionary news for those agents. Much better is the continued survival of the host through the disease-causing organisms exhibiting lower virulence. In this event the host continues to provide the environment within which the disease agent can continue to thrive, but to thrive somewhat less vigorously because they pursue their self-interest less vigorously. In effect, the good of the individual is reigned back for the good of the group. The reader will at once recognize that this is another example of the altruism problem. A reduction in the virulence of some of the agents in the group, perhaps all of them, is a loss of fitness relative to agents that maintain higher levels of virulence. However, as a result the group survives because the host survives.

Two points must be made about this example. The first is that there are data to support the evolution of reduced virulence. The best-known example comes from one of the most famous cases of attempted pest control by a government, the myxomatosis let free in Australia in the hope that it would halt the explosion in the rabbit population. The agent was a virus called *Myxoma*, which, on first being introduced, proved devastatingly infectious and effective in killing rabbits. But after some years the virus was found to be less effective in achieving this end. This was for two reasons. One was that the rabbits had evolved increased resistance to the virus, the other was that, as demonstrated in tests on laboratory rabbits, the virus had evolved reduced virulence. This looks like strong evidence for the operation of group selection, each host rabbit constituting the boundaries for a group of *Myxoma* viruses. Other instances of reduced virulence have been recorded.

The second point is that when group selection occurs, as in cases of reduced virulence, it does so under specific conditions. It is certainly the case that if the benefits of individual selection within the group far exceed the benefits of group selection, then the latter does not occur. Group-level benefit must be sufficiently great relative to individual-level benefit. That alone, however, is not enough if the group remains isolated – a self-contained reproducing population – because reduced virulence is more costly within the group than increased virulence, and hence the more virulent agents eventually do come to dominate and overcome the less virulent. The game is lost to the more virulent agents. How, then, is it possible for group selection to be maintained?

The answer lies in the models that Wilson and his colleagues have developed in which the inevitable eventual extinction of the less virulent agents within groups, the altruists, is balanced by the greater overall reproductive success of the groups that have a sufficient concentration of altruists, *combined* with the periodic dispersal of individual members of each group and the reconstitution of new groups, some of which contain the right balance of more virulent and less virulent agents. The simple, basic model Wilson builds on is counterintuitive until one realizes it is the numbers of each type of agent in the population of groups as a whole that matter, not just what is happening within each group. Consider the simplest case, of a population divided into two groups of equal size but with different proportions of altruists (low virulence) and non-altruists (high virulence). Assigning relative fitness values to each type of agent, what Sober and Wilson demonstrate with easy-to-understand calculations is that while the non-altruists inevitably gain in number within each group, because the group with the higher proportion of altruists produces more offspring overall than the group with the higher proportion of non-altruists, the total number of altruists can be maintained across the population as a whole. However, if the groups are not periodically broken up and reconstituted as new groups, the non-altruists eventually come to dominate in both groups.

In short, group selection occurs only if there is more than one group (this has always been recognized); if the groups vary in the proportion of altruists to non-altruists; if the groups with more altruists grow more rapidly than groups with fewer altruists or no altruists at all; and if the groups periodically fragment and form new combinations, with appropriate differences in the proportion of altruists to non-altruists, which then demonstrate anew the relationship within the group between numbers of altruists and growth.

There has to be a *balance* between the opposing forces of individual and group selection. In Sober and Wilson's words, 'the differential fitness of groups (the force favouring the altruists) must be strong enough to counter the differential fitness of individuals within groups (the force favouring the selfish types)'. This reasoning on the balance of group- versus individual-selection forces must hold in any situation in which a biological entity can be seen as a group made up of component parts. Consider again the relative costs and benefits of group living to a flock of birds. Assume that all the members of the flock start off dividing their time equally between scanning for predators and eating. Individual selection will drive some birds to cheat on their fellows by giving up their obligation to scan in favour of increasing their eating time. However, as the proportion of selfish non-scanners increases, so the overall vigilance of the group declines and the risk of predation for all members rises. If the proportion of non-scanners increases beyond a certain point, all the birds are at risk. The group can tolerate a number of non-altruists but becomes unviable when that number is too high. The benefits to selfish individuals of eating when they should be scanning must be balanced by the benefits to all of shared scanning.

The same scenario applies to the functioning of single whole organisms made up of component organs, cells and the genes the cells contain. Assuming limited energy supplies, the body divides its resources among the component organs, each of which is essential to the overall functioning of the organism. Put crudely, if the liver competes too successfully in the production of liver cells at the expense of, say, the production of lung cells by the lungs, the within-group (i.e. within-organism) success of an individual organ, the liver, is bought at the price of threatening respiration, hence the successful functioning of the group as a whole (i.e. the entire organism).

As Sober and Wilson are at pains to point out, this understanding of the need for nature to strike a balance between individual and group selection in a biological world that is hierarchically structured (the hierarchy extending from genes and single cells, through whole organisms, to groups of organisms) has long been understood and applied to examples as diverse as sex ratios and insect societies. Even those like Hamilton and G. C. Williams, habitually associated with hard-line gene-centred approaches to evolution, have shared in, and advanced, this understanding.

There is a further point to note. None of this detracts from the notion of the centrality of the gene as the replicator in biological evolution. The properties of groups that bestow advantages on individual group members are adaptations that exist at the level of the group. They are no different in kind from adaptations that exist at the level of the individual. Williams and others have long recognized that natural selection results in adaptations at multiple levels of the structural hierarchy, which is what the interactor almost always is. Group-level adaptations are no more or less extraordinary than individual-level adaptations, and group selection is a property of the interactor, not of the replicators. Groups, like the individuals of which they are composed and the genotypes of those individuals, are transient and ephemeral within the evolutionary scheme of things. They do not endure the way replicators endure. Nonetheless, interactors, whether organs, whole organisms or groups, are complex and no less wondrous than the replicators with which they 'cooperate' to maintain the continuity of living things.

Wilson and his colleagues have an interesting way of looking at group selection and the evolution of higher-level interactors. Referencing Szathmary and Maynard Smith, mentioned in the first chapter, who have written of the history of life as a succession of major transitions in which lower-level entities, such as genes, have coalesced into higher-order entities, such as chromosomes, single-celled creatures have come together as multiple-celled organisms, and single organisms have formed social groups, a process culminating in the emergence of culture, Wilson invokes the notion of 'social control' rather than altruism. Social control is the evolution at higher levels of the structural hierarchy of means of constraining the activities of lower-level entities precisely to allow the emergence of those higher-order levels of organization. It is a neat substitution of terminology, because although the words change nothing in the way people think of the evolution of group selection, it removes the connotations of altruism, which in biology have become steeped in genes, and shifts the focus from individual and inclusive fitness, thus allowing one to see groups and culture within this broader biological framework. In the words of another American evolutionist, Richard Michod: 'Cooperation is a critical factor in the emergence of new units of selection precisely because it trades fitness at the lower level (its costs) for increased fitness at the group level (its benefits). In this way cooperation can create new levels of fitness.' Such cooperation is policed, in Wilson's vision, through the social controls of the higher-order entities. This is a law that governs all transitions in evolution from lower to higher orders.

But look here, some might ask, have you not overegged this particular pudding by implying that groups are to be seen in the same way as organisms? Surely it is too much to ask that groups be understood as organisms. The answer to this is both yes and no. Most groups lack the integrated wholeness of single organisms, even though the same balance between selection forces operates at every level of the structural hierarchy. In the context of the Szathmary and Maynard Smith view, this is none too surprising. Multicellular organisms first appeared between 600 and 700 million years ago, whereas the insect societies that are the best candidates for being considered as single organisms, or superorganisms, in the word of American entomolgist William Morton Wheeler, have been in existence for only about a sixth of that time – around 100 million years. Five or six hundred million years of evolutionary time can make a huge difference in functional integration. Nonetheless, experts on insect behaviour like Thomas Seeley of Cornell University paint a picture of social insect colonies which are striking in the degree of integration they reveal. A typical honey-bee colony comprises some 20,000 workers and a single mother queen that has mated with about ten males. With ten or so patrilines in a colony, the workers in general are less genetically related than is often assumed. Nonetheless, because the workers almost never lay eggs of their own as long as the queen is present in the colony, the queen is the 'reproductive bottleneck' through which every individual's gene propagation must pass. For this reason the reproductive success and welfare of the queen is of paramount importance to the colony. Workers shape the cells in which eggs are laid, inspect and clean cells containing eggs, maintain the internal temperature of the hive, store the nectar that foragers bring to the hive, patrol and forage for food and new nest sites, and, of course, defend the colony against intruders – all through coordinated group action the outcome of which are many functions that could never be attained by the actions of a single bee. The behaviour of individual workers is coordinated by a multitude of cues and signals diffused between individual bees and not coordinated by any single individual or small group of individuals. There is no centralized control for any of these activities and functions. Honey-bee colonies may not have the structural complexity of the individual bees in them, but they have a functional complexity and integration that, it is not too fanciful to say, approximate to the functional complexity and integration of each individual bee.

This brings us to the second answer to the question of whether groups of organisms ever equate with single organisms – the one that focuses on part–whole relationships in functional organization. Again, for clarity, the argument will be presented by comparing just groups with individuals, but, as with the notions of social control or the conditions necessary for the evolution of all higher levels, it can be made when comparing any adjacent levels in an interactor that is hierarchically structured. Take the function of any organ in the body of a bee, such as the tracheal system, which serves the combined functions of mammalian blood and lungs (i.e. getting oxygen to body tissues). An intricate structure made of tens of thousands of individual cells, the tracheal system is something that is not present in any one of those cells. Individual bees also have alimentary tracts and eyes, which again are organs whose functions are not to be found in the individual cells of which they are made. In the same way, the ventilation and temperature control of the colony is a property of the collective activity of many bees, not of any single bee. The astonishing effectiveness with which a colony can track the fairly rapidly changing nectar resources as patches of flowers bloom and then die within a kilometre or two of the hive is a property of multiple foragers, not just one animal. The nutritional state of the hive is communicated to foragers by the time it takes for their nectar to be offloaded by food-storer bees, but that time is dependent upon the activity of many food-storers. In short, in individuals as well as in groups, each level of the interactor hierarchy has functional characteristics that are not present at the level below. Groups may have properties that are either simply not present, or, if present, have different functional value, in the individuals of which they are made up.

In their book, Sober and Wilson mention the interesting case of egg-laying by hens. Hens, we all know, have been intensively artificially selected to be productive egg-laying machines, and in recent years have mostly been made to carry out this function in so-called factory farms, where they are punishingly crowded together. One result is high levels of stress in the birds and accompanying high levels of aggression, making it necessary to trim their beaks so as to reduce the injuries they inflict on each other. Hens, however, are, as Sober and Wilson point out, social animals that normally lay eggs in social groups. Selecting for egg-laying by the individual may result in animals that are productive for several reasons, only one of which is being a good egg-layer. The others have to do with inhibiting the egg-laying of others or being resistant to such social forces as aggression. Selecting for egg-laying within a group-selection paradigm, i.e. selecting for females from the most egg-productive groups, has been shown to increase egg productivity very significantly within just a few generations. The effect of group selection is not to single out the isolated trait of egg-laying ability, but a host of social behaviours which reduce stress and aggression levels and hence facilitate the egg-laying of all the hens in the group. This group-level set of social behavioural traits can only be manifested within the context of the group. Take the group environment away and these are functional properties that have no existence.

Just as individual nerve cells do not have the properties of learning and memory that are characteristic of the collective of nerve cells that we call the brains of intelligent animals like bees, and just as thermoregulation of the hive is a property of a colony of bees and not of any individual bee, so any one human being on her or his own does not have the properties of culture, which exists in virtue of the functional organization of its constituent individuals. Of course brain cells have properties – evolved properties – such as cell membranes that shunt ions differentially across their surface and maintain unequal concentrations of ions on either side of the membrane, which is what forms the basis of the nerve impulse, or action potential in the jargon of neurophysiology, and it is these properties that enable individual nerve cells to behave in the ways that are essential for collectives of nerve cells to have the properties that they do. In exactly the same way, each individual human has evolved mind/brain mechanisms, such as Theory of Mind and the ability to form higher-order knowledge states like schemata, which are at once the defining features of our species and the mechanisms that are essential to collectives of humans, i.e. cultures. The parallels between levels are exact. The relationship of the properties of individual nerve cells to collectives of nerve cells in the form of neural networks is exactly the same as the relationship of individual mechanisms of language or Theory of Mind, which, of course, are neural networks, to culture, the collective of knowledge and value.

Taking this same view of functional organization as something that only exists at certain levels of the hierarchy of interactors, and turning attention back to the way in which living in groups has been traditionally understood over the last 40 or so years, makes us revise the way we think about the advantages and disadvantages of living in groups. Some advantages, such as the dilution of danger, are of benefit only in a group-living situation. Solitaries cannot reduce their domain of danger. The same applies to any information transmission, which can only occur within the group and by virtue of the existence of the group. The elegant models which trade-off feeding time and scanning time because of the increased vigilance provided by many pairs of eyes, and which have received straightforward for-the-good-of-the-individual treatment, must really be seen and understood as models of advantages that arise from group-level functional organization.

In many ways, taking a group-level perspective does not change very much. Yes, we can look at the cost-benefit analyses of social groups that appear in standard behavioural ecology texts and see them in a somewhat different light. But often the shift in perspective is slight, and that, perhaps, accounts in part for the reaction of many biologists to the prominent reappearance of group selection as a force in evolution – a shrug of the shoulders. As Sober and Wilson document so well, there is nothing much new here. On the other hand, the perspective of a structural hierarchy, with a balance of checks and advantages across levels, and the existence of functional organization unique to specific levels, provides the interactor with an appropriately complex architecture, which was missing when everything was lumped into the vehicle as some single entity.

But it does more than just that. It provides us with an evolutionary framework within which we can begin to understand how the group-level property of first culture could have been selected for, and how the properties of human intelligence could have been further changed by the evolutionary forces of that selection of a group-level property, resulting in the contemporary human mind and its capacity for entering into modern culture. It also gives us the proper biological architecture within which to place culture. Whatever else culture is, it exists only in virtue of a social group. It is a property of the human interactor, which is a very complex interactor indeed.

There is another, and somewhat different, way of approaching group selection from that taken by Sober and Wilson. In their 1985 book *Culture and the Evolutionary Process*, Robert Boyd and Peter Richerson develop a frequency-dependent bias model to account specifically for human group selection and the evolution of cooperation. At the heart of this is the notion that humans have a *genetic* predisposition to choose between *cultural* variants, be these beliefs, attitudes, values, specific overt behaviours or anything else that can be culturally transmitted, on the basis of the frequency with which they occur in a group or population. This bias must not simply reflect the relative frequency of occurrence of a cultural variant. Rather the naive individual must be disproportionately likely to choose and act on the variant which is more common. Psychologically this is an entirely plausible assumption, as the review on social force in chapter 5 demonstrated. There is a great deal of evidence to support the idea that most of us are powerfully affected by majority actions and opinions – not everyone, and not always, but most humans are inclined to conform and often enough we do so. Whether this responsiveness to social force is in part genetically caused is simply unknown at present, but Boyd and Richerson are biologists writing a theory of gene–culture co-evolution, so, as we saw in chapter 4, the assumption of a genetic part-cause for some psychological dispositions is normal and necessary to such modelling. It is a hypothesis the truth of which will be tested as the genetic basis of brain and psychological mechanisms becomes known over the coming decades. In any event, Boyd and Richerson show how a conformist frequency-dependent bias in their models provides a simple, if general, rule that leads to group-level selection of cooperative cultural variants that, even though they may benefit the group at some expense to the individuals that make up the group, may be favoured by natural selection and hence may evolve 'because it increases the chance of acquiring locally adaptive cultural variants in a heterogeneous environment'.

Their models, like those of Sober and Wilson, make additional assumptions about migration between groups and the extinction and reforming of groups. And, like Sober and Wilson, they are aware of the need for group-level forces to counterbalance the strength of individual selection, and how tension between group-level and individual selection is the basis for all social dilemmas – the interests of the individual versus the interests of the group. The principal difference is that Boyd and Richerson posit a specific human psychological trait as the basis for group-level selection, whereas for Sober and Wilson the group is located within a complexly structured conception of the interactor that applies much more widely than to just our species. Of course, neither account excludes the other. A complex interactor comprising particular psychological mechanisms that allowed for the emergence of first culture and its subsequent evolution into contemporary culture does not exclude from these mechanisms something like frequency-dependent bias in the response of humans to cultural variants. Once again Occam's Razor does not necessarily rule. A pluralist multideterminism might be the appropriate way in which to cast a science of culture.

Niche construction

Why, one may ask, all this fancy biology? What is the necessity for formal models with numbers attached, talk of reduced virulence or unequal sex ratios, and images of differing functional organization in a structural hierarchy? Surely it requires nothing more than plain commonsense to understand that once human cognition had reached the point at which each individual could calculate, at least to some approximation, what the costs and benefits were of acting collectively, in, say, defending oneself against the actions of outsiders, or in gaining access to a resource inaccessible to individuals working alone; once intelligence had reached the understanding of how useful such devices as singing or dancing were for strengthening group solidarity under such circumstances, hence was capable of thinking up ways of improving them, such as rewarding with status or material goods those whose service to the group was outstanding; once human reasoning had reached the point of seeing how trading off the relative costs and benefits of what was good for the group and what was good for the individual led inevitably towards the social dilemmas and their solutions that are the stuff of human ethics and morality; why, once humans had come to such understanding, surely the advantages of collective knowledge, skills, beliefs and values would have been obvious, and, when circumstances dictated, people would have acted in concert. We surely do not need science to tell us what is obvious.

Well, that science should not tell us what is already known or within the compass of ordinary understanding is certainly correct. But the argument of the last paragraph is not science, hence won't do as an account of the evolution of culture. It is 'folk culture' talk, analogous to folk psychology. The latter is the commonsense knowledge of 'what makes people tick' which derives from deeply rooted, evolved cognitive modules that give us the ability to understand others, but equally gives us no insight into the causal processes and mechanisms that make each and every one of us natural psychologists. It is the understanding of those causes that defines psychology as a science. In the same way as we have this intuitive knowledge of how people work, and indeed of physics and the animate world at large, so we have intuitive knowledge of culture. All of us clearly have some understanding of the potency of identifying with groups, of being accepted by others, and of the power of collective action. Such understanding does not, however, reveal the conditions under which culture evolved or what the causal processes and mechanisms that give rise to culture are. That is why we have to place understanding within a firm biological and psychological setting. Thus it is that we have one final set of biological ideas within which to locate human culture.

At the centre of these is the notion that organisms, roughly speaking Hull's interactors, change the world in which they exist. At first glance this may appear to be a trite conception. We know, for example, that for close on two billion years the atmosphere of Earth was without oxygen. It was the evolution of cyanobacteria – some of whose photo-synthetic pathways liberated large quantities of oxygen, a portion of which was converted into ozone, which, in forming a protective shield against ultraviolet radiation, had a profound effect on the subsequent evolution of life – that provided the first step in a well-understood sequence in the history of life on our planet. The idea, then, is hardly a new one. The events leading to the appearance of oxygen in our atmosphere, however, were not generalized into any kind of evolution-ary principle concerning the way organisms may generate the con-ditions that influence the selection forces acting upon them. As so often, it is to Darwin, that most fecund of biologists, who seemed to have thought about everything, that the original idea can be traced. Waddington, nearly a century after Darwin, revived the idea by posit-ing what he called an 'exploitive system', which he deemed as impor-tant as natural selection. Waddington's exploitive system was the collection of all possible ways in which the behaviour of organisms, including the 'choices' of animals with complex cognitive capacities, modifies the selection pressures acting upon them. Subsequently the idea appeared in a variety of forms and for various reasons. Alan Wilson, the American geneticist, and others developed the idea of 'behavioural drive', whereby the behaviours of animals have an effect on the pace of evolution, largely on the basis of data pointing to a relationship between brain size and rates of anatomical change in species. Ernst Mayr, perhaps the greatest living evolutionist, wrote in 1982 that 'many if not most acquisitions of new structures in the course of evolution can be ascribed to selection forces exerted by newly acquired behaviours'. And in an influential sequence of publications Richard Lewontin, reacting to what he saw as the excessive passivity in the way adaptations were being increasingly thought of, advocated a more constructive approach to the understanding of adaptations specifically and evolution generally. Organisms *do* things, and that doing is often significant. It might be noted that Dawkins' extended phenotype is an idea not a million miles from this line of thinking.

In recent years it is in the work of John Odling-Smee and his colleagues that these ideas have been brought into sharpest focus, in the guise of 'niche construction'. Take the simplest case of the shade cast by trees and the leaves they shed. Both shade and shed leaves have significant effects on surrounding vegetation and the subsequent competition for light, water and nutritional sources. Trees create conditions in their immediate environments which have significant consequences for their survival and vigour. Well, that is a very passive case, and hence perhaps none too impressive as an example of how living things can alter the circumstances of their lives. Consider, then, the large number of mammalian species that construct elaborate burrows, with multiple entrances and extended underground passages connecting chambers of varying sizes. These underground worlds are cases of literal niche construction because they provide protection from enemies and the elements. They most certainly also provide a change in selection pressures acting on the animals that make them, and there is evidence that burrows started out in geological time as relatively simple pits which had sufficiently constant effects on selection across generations that they led to the further evolution of burrow construction and of burrowing behaviours, such as defence and maintenance, resulting in the kinds of complex burrow systems and the behaviours attendant upon them that we see in some species, like rabbits and mole rats, today. This legacy of a constructed niche with its modified selection pressures is referred to by Odling-Smee as 'ecological inheritance'. The important point to note is that these niches must be stable across multiple generations, for it is in such stability that they provide the consistent selection pressures that become the forces of evolutionary change – selection pressures that are a consequence of the behaviour of individual animals. This is truly a case of the animal phenotype as interactor or operator. And burrowing mammals are not isolated examples. One could cite innumerable other species and their behaviour, ranging from insects that provision their eggs with a food supply, to earthworms, the behaviour of which is so beneficial to the chemistry and structure of soil (and which gave Darwin the idea in the first place) as instances of niche construction.

Now, digging a burrow or laying one's eggs in the nest of another species of bird, as does the cuckoo, requires no powers of intelligence. These are innate behaviours, instincts, and conventional evolutionary theory needs only to be tweaked to take into account this additional feedback loop whereby particular behaviours, part-caused by genes, become forces for further evolutionary change. Add behaviours that are generated by intelligence and one has a second tier in this particular hierarchy, a hierarchy of heuristics (see chapters 2 and 3), but with the same dynamic of the consequences of intelligent behaviour feeding back via consistent selection pressures created by that behaviour, to add a further feedback loop to the evolutionary process. One example of the way intelligent behaviour can have evolutionary effects is how learning to grub with a tool has altered the selection pressures acting on the beak of the woodpecker finch.

It must already be clear to the reader where this is heading. Add the third heuristic of culture to this hierarchy of knowledge-gaining processes. Now the feedback loop originates in culturally caused changes in the world, be these structural artefacts like clothes and shelters, inherited bodies of knowledge such as science and medicine, or persistent value and belief systems such as the preferred sex of children because of rules of inheritance, dowry customs or plain labour requirements, or who is allowed to enter into marriage with whom and hence what mate choice is allowable – and culture never has greater consequence for human biology than when it determines mate choice. Well, whichever it might be, the alleviating of the effects of climate, the treatment of illness, or choices determining which children are born and which survive and thrive, in every case what culture must be doing is constructing niches, both physical and cultural, that persist across generations and alter the selection pressures acting on the people who make up that culture. In this case, ecological inheritance becomes cultural inheritance, and culture becomes a powerful force in human evolution.

The niche-construction conception provides the architecture of a hierarchy of heuristics, the knowledge-gaining processes of the main evolutionary programme and the evolved subsidiary heuristics of individual intelligence and culture, with a further framework for integrating the social and the biological sciences. It allows us to understand culture as a form of intensified niche construction and ecological inheritance, in which the niche extends deep into the individual mind and binds it tightly to the minds of others and to the material world of culture. Culture can now be seen as a natural extension and expression of processes that are actually billions of years old.

Suggested Readings

Dawkins, R. (1982) *The Extended Phenotype: The Gene as the Unit of Selection.* Oxford, Freeman. (A robust defence of gene-centred evolutionary theory.)

Krebs, J. R., and Davies, N. B. (1995, 3rd edition) *An Introduction to Behavioural Ecology.* Oxford, Blackwell. (A crystal-clear account of altruism, the ecology of group living, and much else besides.)

Laland, K. L., Odling-Smee, J., and Feldman, M. W. (2000) 'Niche construction, biological evolution, and cultural change.' *The Behavioural and Brain Sciences*, vol. 23, 131–75. (A recent review of niche construction, its application to intelligence and culture, and comments on it by biologists, psychologists and anthropologists.)

Michod, R. E. (1999) *Darwinian Dynamics: Evolutionary Transitions in Fitness and Individuality.* Princeton, Princeton University Press. (A technical account of a unifying theory of evolutionary transitions, including the evolution of individuals and groups.)

Seeley, T. D. (1989) 'The honey bee colony as a superorganism.' *American Scientist*, vol. 77, 546–53. (Book-length accounts by the same author are also available.)

Skyrms, B. (1996) *Evolution of the Social Contract*. Cambridge, Cambridge University Press. (A readable introduction to game theory.)

Sober, E., and Wilson, D. S. (1998) *Unto Others: The Evolution and Psychology of Unselfish Behaviour*. Cambridge, Mass., Harvard University Press. (An account of group-level selection and its consequences.)

7

The Strangeness of Culture

In a recent collection of essays, the eminent American cultural anthropologist Clifford Geertz provides a clear example of what might be called the great cultural anthropological fallacy, compounded by a *non sequitur*, which together lead to a bleak form of anthropological isolationism that likes to pass as an interesting pluralism. Geertz's central conceptual plank is that diversity is the essence of our species and that it is this diversity that the social sciences should be trying to understand. Well, this is an arguable, if defensible, position. Certainly human diversity in the sense of cultural diversity must be explained, whatever scientific direction one is coming from. But the truth of diversity then forms the base for making the illogical leap to the fallacious claim that there is no common core to the human mind, no 'mind for all cultures', which is what allows all humans to be creatures of culture. Geertz is not denying that there are universal features of human psychology, such as learning and the ability to communicate, that in some sense are implicated in the human capacity for culture. What he is saying is that the implication of these features, hence their causal force and their explanatory power, is trivial. This, it will be remembered from chapter 3, is the notion of culture as a 'level' of phenomena pretty well divorced from all else.

Now Geertz is not alone in this view. His is simply the most prominent of the voices that speak of the supposedly unbridgeable gap between culture and other human traits – or, if bridgeable, only in the most banal and empty manner. What drives intelligent and knowledge-able people to the empty conceptual view that culture cannot be linked in significant ways with other human attributes, especially those of our psychology? Certainly it is a view that was not shared by great anthropologists of the past like Margaret Mead. Part of the answer, as indicated on several earlier occasions in this book, lies in the destructive oversimplification that some biological approaches, perhaps all, have in the past wreaked on the anthropologists' view of culture as something at once immensely complex and extremely subtle. Another and related part of the answer is the perception that no biological approach to culture has yet even hinted at a possible explanation of that perceived complexity and subtlety. And yet another part of the answer lies in a strange quality of culture itself – a strangeness that could be mistaken for magic. Well, culture might have strange qualities, be different from other human traits and perhaps be unexpected in some ways, but whatever else it is, as will be recalled from chapter 3, it is not magic. It simply must be understood in the context of accepted processes and mechanisms. The strange quality is best thought of as an 'added value' or 'added force' inherent in what culture is and what it does. By this is meant that culture has a feel to it of something that goes beyond the individuals of whom it is made up – of embodying the cliché of being more than the sum of its parts. This should not be taken to represent some kind of cultural dualism analogous to Cartesian dualism. The French philosopher René Descartes postulated the existence of *res cogitans*, a non-material, non-extensive mental world unique to humans, to be distinguished from *res extensa*, the physical self that has substance that can be measured in various ways. This is the classic mind–body dualist position. No one is suggesting a form of immaterial cultural *res cogitans*. Nor is it simply a kind of illusion that arises from culture's embodiment in artefacts like tools and buildings, and its extension across generations, and hence its seeming lack of attachment to any one person or even group of people. Indeed, it is almost the reverse. Social constructions, such as baaskap, justice and money, are *things that come into existence only because of culture* and do not necessarily have physical or temporal extension (though they usually do). Of course, it is clear that other elements of culture, such as higher-order knowledge structures like schemata and scripts, are also products of culture, but while these are undoubtedly significantly shaped and formed by culture, in principle they could exist within the mind/brains of individuals who are not a part of culture and may well do so in animals that have no culture. This is not the case for social constructions. Whatever these are, they exist only in virtue of cultures. This is why chapter 1 picked out social constructions for special treatment. They have an odd 'feel' to them because they cannot exist in only a single mind. They are born of multiple minds. That, however, is not the only source of their strangeness.

Any kind of thought held in my own head and shared with nobody else is an entity entirely within my own mental sphere. When I share it with others, then whether it assumes this quality of extra force or extra value depends on the nature of the thought. If what I am sharing is a reference to the tangible, material, everyday world – this mountain or that person – the thought in the heads of two or more people is little changed. But if the thought is an invention that goes beyond the material world of reference – the mountain where the dead dwell, or this person of honour – then sharing the invented quality – of the dwelling dead, or honour – adds to belief in what is not material and tangible. The ability to have thoughts, including values and beliefs, that go beyond the world of our senses is, perhaps, the one truly unique human quality, the one thing that makes us different from all other animals. Turning those invented qualities into forms of reality is what culture achieves. In this sense, culture is indeed more than the sum of its parts. Culture is an imagined world made real. That is why it seems strange. To say this is not to make any fundamental departure from the ideational approach that culture is shared knowledge and beliefs. It is merely to recognize that what is learned from others is, in large part, created and imagined knowledge. This is not to deny that learning to use a knife and fork is one very simple form of culture. But it does focus attention on what is much the greater part of all human cultures. This is knowledge of created worlds and of values and beliefs based on those creations. Art, deities and money are cultural *inventions*. They did not exist prior to human culture, and they do not exist outside of it. And it is precisely because they are constructions of our collective minds that we need to understand their existence in terms, as it happens, of a relatively simple psychological mechanism. Belief in the reality of invention only comes with numbers. We need more than one mind for that. Hence the attention given to group-level selection in the previous chapter and to social force in the chapter before that. Hence also the assertion in chapter 1 that if, and only if, one can get a biological bite on social constructions, then and only then can one claim the beginnings of a biological understanding of the complexity of culture. Thus it is that this chapter returns to where this book began. To continue the metaphor of the preface, the threads of this final chapter are mainly social, though we will try to see how they interweave with the biological threads of previous chapters.

The construction of social reality

The subhead is the title of an important book by John Searle, and I make no apology for devoting this section entirely to this work. In his book Searle provides an analysis of the strange quality of culture, specifically in the form of what we are calling social constructions, things that exist only because people agree that they exist. It is a philosophical analysis that makes little reference to psychological processes and mechanisms. However, Searle is wonderfully adept at laying out the problem, and so provides us with a substantial starting position for trying to understand that mysterious quality of culture. Science of any sort will enter the argument in subsequent sections.

Searle rests his analysis on two features of reality, of what *is*, that he considers to be not optional in any understanding of ourselves and the world we live in. The one is represented by the atomic theory of matter. Everything comprises particles or aggregates of particles (the strings and superstrings of theoretical physics can count as particles in this most general sense), be these fundamental particles like electrons or neutrinos, force particles like photons or bosons, or their aggregation as elements, compounds, mountains or orang-utans. The second is encompassed by the theory of evolution, which accounts for how, through the action of natural selection and other forces, cellular and organ structures have evolved, including the human brain, which causes and maintains consciousness. With consciousness comes intentionality, which Searle defines broadly as 'the capacity of the mind to represent objects and states of affairs in the world' apart from itself. This includes intentional states such as knowing, believing and wanting both in oneself and others.

Consciousness is a biological, and hence physical, property of humans, but its intentional properties allow a fundamental distinction to be drawn. This is the distinction between 'brute' facts, such as the presence of snow and ice near the summit of Mount Everest, and 'institutional' facts, which are dependent upon human intentionality and agreement. The state of being married is one such institutional fact, a socially constructed reality based on a whole set of agreements between people about a variety of matters, ranging from entry into a contract of marriage, including any rituals marking that act, to rights and privileges relating to property, income, children and such like, which are the consequences of entering into the contract. Now brute facts are totally independent of human opinion. If humans did not exist, or existed but were not conscious, there would still be snow and ice on Mount Everest. (True, brute facts require the human facility of language for their stating, and language is a particularly potent form of social fact. But that merely forces the further distinction between the fact stated and the statement of the fact. Mount Everest's snow and ice exists whether we can talk of it or not.) Brute facts are intrinsic to nature. Institutional facts, however, are wholly dependent not just on humans but on human agreement. If we did not exist, or if we did not agree on the state and consequences of marriage, marriage would not exist. Institutional facts, unlike brute facts, are intrinsic to human intentionality.

This distinction is absolutely crucial to Searle's analysis, and he extends it into intrinsic and observer-relative features of the world. He does this using the example of an object made partly of wood and partly of metal, these being the object's intrinsic features. That the object is also a screwdriver is an observer-relative feature of it. It is a screwdriver because the person who constructs it, and those who use it to drive screws, are able to represent it as, and use it, for that purpose. The reality of the object as a brute fact of wood and metal is unchanged: what has been added is an epistemic feature, which is relative to its observers and users. This epistemic addition is a result of certain characteristics of the observers and users, of course, which characteristics are themselves intrinsic features of those observers and users – which is to say, in natural science terms, that these intrinsic features are psychological processes and mechanisms arising from the structures and functions of the observers' and users' brains.

How, though, are we to understand observer-relative features as they relate to the institutional facts of social reality? How do they give rise to social reality? Searle answers this with an argument for three essential elements. The first, the assignment of function, he considers a particular capacity of human intentionality, though he allows the possibility of some rudimentary form of it in certain other species of animal. We assign functions to natural objects (trees function to cast shade and rivers are for washing in), and we also construct objects – artefacts – to fulfil certain functions (such as screwdrivers, chairs and houses). Searle, somewhat controversially, insists that functions are never intrinsic to objects but are always observer-relative. (Nature, he asserts, is a very long string of causal events. The heart is so constructed that it causes blood to be pumped: it takes a human being with a specific set of values and other mental traits to claim that it is the function of the heart to pump blood. 'As far as nature is concerned intrinsically, there are no functional facts beyond causal facts.') Functions are assigned, or imposed, relative to the minds of users and observers. There is a further distinction to be observed (this is, after all, a philosophical analysis, though I am playing rather fast and loose with Searle's distinctions and for simplicity have left some of them out). Agentive functions have specific purposes – this is a chair and that is a screwdriver. Non-agentive functions are those we impose on nature – the function of the heart is to pump blood – but which do not serve our intentional goals.

There is, says Searle, one very important class of agentive functions: the functions of those things that are used to stand for something else. Standing for something else is what those things do and what their purpose is. When, as an aid to guiding someone to my house, I draw a map, which as an object, a brute fact, is merely a series of pencil marks on a piece of paper or scratches on a bed of sand, that map has the agentive function of standing for, of representing, a spatial array of objects like streets and landmark buildings. I have imposed an intentional function on the paper and pencil marks, or the sand and scratches, which are not intrinsically intentional. They are intentional only relative to myself or whoever reads the map. This is the intentional function of meaning or symbolism. Language, whether spoken, signed or written, is the supreme instance of a social function of meaning imposed on the brute facts of sounds, movements of vocal tracts or limbs, and marks.

The second essential element in how observer-relative features relate to social facts is what Searle, with a very nice turn of phrase, calls 'collective intentionality'. Collective intentionality, quite simply, is shared intentional states. The 11 players in a football team share the goal of winning the game by way of shared knowledge of team tactics, set plays and the weaknesses of their opponents. They also, of course, share an identity by being a part of this or that club. Collective intentionality is not reducible to the sum of the individual intentionalities of the 11 team members. This is because collective intentionality may give rise to individual intentionality. The latter can stand on its own, of course. But when collective intentionality exists, it does so prior to the intentionality of individuals and may be an additional source of individual intentionality. When a winger crosses the ball it is the result of that player's intentional states, but standing behind those and giving rise to them is the collective intentionality of the entire team's coordinated knowledge that the opposition is weak when defending against crosses. By the same token, when two people are fighting one another in a boxing ring, they share the intention to fight – their 'we-intention' is that of taking part in a boxing contest – but this is entirely different from the individual 'I-intentions' of the participants in a bar-room brawl.

Searle is very determined on this point because it lies at the heart of his analysis and, as we shall see, it is also absolutely central to psychological and sociological analyses of social behaviour. It is the very source of culture's seemingly magical quality of 'added force'. For Searle, collective intentionality is a 'primitive phenomenon', by which I take him to mean that it is the product of some fundamental psychological trait common to all humans. Contrary to the views of some other philosophers, Searle's collective intentionality – to repeat the point, because it really is very important – is not the sum of individual intentionalities and is not reducible to them. It is something different. This does not mean that collective intentionality is some absurd form of collective consciousness that hovers in the spaces between people or binds their minds through mysterious field forces. Each of us has a mind and mental life that are confined to our individual brains. But, to quote Searle, 'it does not follow from that that all my mental life must be expressed in the form of a singular noun phrase referring to me. The form that my collective intentionality can take is simply "we intend", "we are doing so and so", and the like. In such cases, I intend only as part of our intending. The intentionality that exists in each individual head has the form "we intend".' And *every* institutional fact involves collective intentionality.

It is impossible, I think, to overstate the importance of collective intentionality. It is not only at the centre of Searle's scheme: it is at the centre of all the social sciences. It is what makes humans human. There is also pleasure in it, which Searle does not pick up on. If we enter into collective intentionality with those with whom we have strong emotional ties, then routine activities like eating a meal are transformed into pleasurable acts. We do things with others, such as going to the theatre or on holiday, which most would not do on their own. And the change from an I-intentional state to a we-intentional state is what people seek when they gather round the water fountain at work for a chat, go on their own to a pub or join a singles club. The social and the emotional are inextricably linked.

The third essential element in how observer-relative features relate to social facts is constitutive rules, which are to be distinguished from regulative rules. The latter regulate existing activities. For example, most people travel on roads and pathways because, among other reasons, this is easier than moving across fields and over hedgerows. But then traffic must be regulated to minimize hold-ups and stoppages. In Britain people drive on the left. In most parts of the world the rule is to drive on the right. Well, it doesn't matter if it's left or right as long as everyone does the same thing and the traffic is regulated. Constitutive rules are different in that they create the conditions for activities like games or parliamentary governance. The rules of football create the activity of football, and the rules of parliamentary election and parliamentary procedures create the activity of government by parliament. Ease of traffic flow does not depend one jot on whether the rule is drive on the right or drive on the left. But change the rules of football by allowing, say, the carrying of the ball in addition to the kicking of it, and you have a whole new ball game.

All institutional facts exist within a framework of constitutive rules, and this framework can be a very complex system indeed. When I go shopping with a credit card, the transaction of buying a book is enmeshed within a system of constitutive rules according to which the card stands for money and money stands for the capacity to acquire a good, and the book-seller relinquishes ownership of the book to me within a particular legal framework of what it means to own something, and yet the actual contents of the book, its copyright, are not mine and were not the seller's but are the property of the publisher and author – to mention just a few of the constitutive rules of book-buying. Others relate to my right to make any kind of transaction, to have a bank account and an appropriate credit rating, and to walk the streets of London rather than being incarcerated in a prison.

The assignment of function, collective intentionality and constitutive rules are the essential elements of Searle's analysis of social reality. In a nutshell, Searle argues that these are the necessary ingredients for the creation of social reality by the collective imposition of functions, functions that can only be performed in virtue of collective agreement and acceptance. At the heart of all social reality and institutional facts is the structure 'X counts as Y in C'. Thus, paper printed in a particular form by the Royal Mint (X) counts as money (Y) in Britain (C). And a specific ceremony performed by a particular type of functionary (X) counts as a marriage contract (Y) in Britain and many other countries (C), but may not be recognized within certain religious frameworks. The conditions of X must be changed in terms of the carrying-out of certain specified rituals under the guidance of an individual of particular background and training if it is to be considered a marriage within, say, Muslim or Jewish traditions (C). That basic structure can be, and usually is, iterated, the degree of iteration being a measure of the complexity of social reality. The structure 'X counts as Y in C' can be repeated for each X and Y term. Thus the officiating individual has the capacity to perform a marriage ceremony only because certain training and study (X) confers specific religious or civic status (Y) in specific countries (C); and a marriage contract (now X) counts as a set of mutual obligations and rights concerning such matters as sharing property (now Y); and property (now another X) stands for a legal claim of ownership of something (now another Y). And so on.

'The central span on the bridge from physics to society,' writes

Searle, who clearly will have no truck with the isolationism of many cultural anthropologists, 'is collective intentionality, and the decisive movement on that bridge in the creation of social reality is the collective intentional imposition of function on entities that cannot perform those functions without that imposition.' The physical substance, the brute fact, of money or land or a written marriage contract is not enough, indeed is often hopelessly inadequate, to perform those functions. The literal material value of a 20 pound note, i.e. the value of the paper and the dyes on it, is ludicrously small and of virtually no consequence at all to the average Londoner. Yet the same note will get me whisked across the city in a taxi, and the driver will then dine out (if cheaply) on that same piece of paper. Similarly, standing on a piece of land, at least in western Europe in the third millennium, does not signify ownership. Indeed, we can own land that is far distant from our physical selves. Ownership is vested in titles to properties usually following appropriate financial exchange.

Importantly, these functions arise out of constitutive rules that require not only collective agreement and acceptance, but *continued* collective agreement and acceptance. X only counts as Y in C provided the agreement holds. When the conditions for agreement break down, so too do the institutional facts. Baaskap and apartheid (white skin (X) counts as a person with rights and privileges greater than those of others (Y) in South Africa (C)) collapsed and ceased to have legal force when collective agreement as to their acceptability was withdrawn, or at least ostensibly withdrawn by the white population of South Africa, and overruled by the counter-agreement of non-white South Africans that baaskap is an unacceptable basis for government. At several times and places in the last century money ceased to have value (in 1920s Germany and early 1990s Bosnia, for example), and other objects, such as packets of cigarettes, took the place of money, or societies so affected reverted to barter.

Well, whether it be ten dollar bills or packets of Marlboro cigarettes, the function imposed on these things through collective intentionality entrapped within networks of constitutive rules is what creates social reality. This function does not derive solely from the physics and chemistry of the objects, from the brute facts of the pieces of dyed paper and the cylinders of rolled tobacco. What it derives from is the neural-network states in the brains of large numbers of individuals being in a particular relationship to one another, and that is an extraordinary thing.

This is only the barest outline of Searle's opening position in what is a richly extended piece of work, but it has done the job that needed doing, which was to present the salient features of social reality. Much of the rest of Searle's analysis is concerned with language, the creation, maintenance and extension of social reality, and a defence of his realist stance, which need not concern us here. Suffice it to say that for Searle language is essential for collective agreement because what is being agreed exists only in virtue of the agreement: 'there is no prelinguistic natural phenomenon there. The Y term creates a status that is additional to the physical features of the X term', and that status exists only because people believe it exists, and the meaning of the belief and the agreement about it can only be cast in symbolic terms. Thus money is much more than the paper or coins of its physical embodiment, and because that status exists not as a brute fact but by convention, there has to be a conventional means of representing it. Language is that conventional means.

The central and essential point of Searle's analysis is that what he nicely refers to as a sort of metaphysical giddiness arises when we ponder institutional facts as social reality, but simply never occurs when we consider physical objects like screwdrivers or mountains. We get giddy because, unlike 'This is a chair so I can sit on it', the statement 'This is money so I can feed myself' involves a non-physical and non-causal relationship between the X and Y terms in the structure 'This paper (X) has value as money (Y).' Well, does X *really* count as Y? Is this piece of paper *really* of any value? Do I *really* own a house? Am I *really* a citizen of the United Kingdom even though I was born thousands of miles away from northwestern Europe? And is the United Kingdom *really* united? I do not have doubts about what I am seeing when I look at a snow-covered mountain, but all social reality can, when examined closely, bring on this giddy sense of wonderment at its *really* being so, because all it hinges on is the sustained agreement of the people making up a group. That is all. That is 'the most mysterious feature of institutional facts', according to Searle, and he is right. It is all the more mysterious that such things as constitute social reality, which have no causal physical substance outside the heads of those individuals who make up a group, nonetheless do have deadly causal force, as pointed out on several earlier occasions in this book. However, sustained collective intentionality is only mysterious if you do not see this most central feature of culture as serving a purpose, as having a function. Remaining true to Searle's position that function is always relative to human intentionality, in this case the intentionality of the writer, the explanation offered for collective intentionality is that it is an evolved group-selected characteristic of humans. The move from I-intentionality to we-intentionality may mark the crucial transition from the kind of culture chimpanzees have, protoculture, to first human culture, preparatory to its subsequent shaping to contemporary culture. And culture, of course, is a group-level adaptation.

A sociological turn

Interest in social reality, in how things social become possible and their impact upon humans, did not have its origins in late-20th-century philosophy. The writings of German philosophers like Karl Marx and Dilthey, from the middle of the 19th century, are usually cited as the principal source of this line of scholarship. Marx, for example, explored how human activity affected human thought and the role of such activity in determining social relations. Dilthey's concern was the impact of history on thought. Both, and subsequent, writers were caught up in trying to understand what the sociologists Peter Berger and Thomas Luckmann call the 'vertigo of relativism'. The relativism they refer to is the differences in the social reality of different social groups, be these groups the Fore people of Papua New Guinea and the citizens of Surbiton in Surrey, or a mining community in northern France and a fashionable social set in Paris.

Broadly defined as sociological, this is an approach that forms a kind of conceptual halfway house between the philosophy of the likes of Searle on one side and social and cultural psychology on the other. I shall use Berger and Luckmann's classic *The Social Construction of Reality* as the vehicle for presenting these ideas. First it must be made clear that despite the title of their book (which is quite different from that of Searle's), and whatever subsequent use their work has been put to, Berger and Luckmann are not rabid postmodernists. While they write within a tradition characterized by the central idea of a 'dialectic' of the social, which is both formative of the mind and yet is itself formed by the minds of individual humans living in social groups, they do not assert an extreme relativist view that all of reality is socially constructed. They cite the sociologist Karl Mannheim's exclusion of mathematics and some of the natural sciences from a sociology of knowledge with approval. What they are after is understanding of ordinary knowledge, everyday knowledge, rather than lofty ideas and ideology, because it is precisely this kind of quotidian knowledge that 'constitutes the fabric of meanings without which no society could exist'. So, it's work, what to eat and how, friendship, and how to be popular that are really at issue here, rather than ideologies revolving round capitalism, the mother country and all that – though doubtless such matters are also encompassed by the authors' approach. A health warning is also in order. This being social science, the jargon can sometimes be painful. Berger and Luckmann do, however, address very important matters of culture that are part of the bridge between the social and other sciences and hence merit our attention.

Berger and Luckmann's starting point is an obvious one. The social world is that part of the world out there that is made by people expressing their thoughts, feelings and needs. In order to get from the subjective – that which is inside our heads – out into the social realm, these thoughts and feelings must be expressed and made public. Berger and Luckmann call this process objectivation. Objectivation can take the form of an action, a facial expression, a grunt or cry, an artefact, or a speech act. The last of these, as with Searle but approached from a somewhat different direction, falls into a very important class of objectivations termed signification – the production of signs the explicit intention of which is to express subjective meaning. Language is much the most important sign system of humans, but it is certainly not the only form of signification. An artefact appropriately used can serve as a sign, as can dance and music. Berger and Luckmann attach significance to the extent to which signs can be detached from immediate subjective intention, from the here-and-now of subjectivity, and emphasize that language is the supreme system of signs in terms of such detachability. It should be added that they quite independently adopt the Geertz position that the essence of being human is to be socioculturally variable, and that this variability is not the consequence of human nature 'in the sense of a biologically fixed substratum'. Well, since they were writing their treatise in the 1960s, they can be forgiven this error. They are slightly closer to some con-temporary psychological views when they note that if humans do have a nature, its significance lies in our ability to construct ourselves – though if this is to be understood as 'I construct myself' rather than as 'We construct each other', this downplays the human susceptibility to cultural forces, a susceptibility that is central to the authors' arguments.

Next in line in Berger and Luckmann's analysis is habitualization and institutionalization, which, terrible though these words are, bring us to the heart of the matter. Habitualization is the process whereby any frequently repeated action assumes an efficient and economical fixed form. The action does not have literally to be an act, it could be a way of thinking or solving a problem or telling a story, but it is most easily conceived of as a relatively simple doing of something. Habitualization is a kind of practice-makes-perfect process that com-pensates for 'man's undirected instinctual structure'; using the term in this sense Berger and Luckmann are in effect saying in their rather odd way that humans have to learn their skills. Institutionalization is what happens when two or more people interact by way of 'reciprocal typification of habitualized actions'. Such typification *is* an institution, which is any group of people interacting in ways that coordinate their skills or their knowledge. Put people together and they form institutions. And in forming institutions, the activity thus insti-tutionalized 'has been subsumed under social control'. It is history that determines just what kinds of institutions characterize a particular social group, but the important claim Berger and Luckmann are making is that the tendency of humans to form institutions is what makes humans human. Humans form groups whose coordinated activities are socially controlled.

But just what does it mean to say that an activity is under social control, and how can institutions be maintained in time? As with

Searle, these questions probe at that strange quality of culture, and Berger and Luckmann's treatment of it, even though it must be given the awful title the 'objectivation sequence', is very neat. Here is how it goes. In the beginning, Robinson Crusoe is alone on the island. Unformed by instincts he adjusts to the requirements of his new life by using his intelligence to adopt a range of habitualized behaviours suited to that new life. Each day, sighing 'Here I go again', he sets about his well-practised routine of replenishing his fresh-water supply, foraging for food, writing up his journal, then trying to build a boat that will return him to his previous life. After Girl Friday appears on the scene, it doesn't take long for the two to become familiar with each other's routine because they are, after all, naturally rather interested in one another. 'There he/she goes again,' each mutters as the one observes the habitualized activities of the other. As it happens she is much better at foraging than he is, and it becomes obvious that if they work together, their combined behaviour of reciprocal typification of habitualized foraging leads to better results than when each searches for food on their own. So they quickly become an institution, and what they now say as they set out each morning is 'Here *we* go again' (Searle's we-intentionality). Since he is the better cook he takes on that duty entire, while she spends the same time updating the journal because she is much the better writer. Thus they adopt specific roles, combine their activities when efficiency demands it, and generally become a stable, predictable, social unit with appropriate division of labour. Innovations, such as his invention of dance to divert and enchant her, and her development in the role of storyteller to charm and entertain him, enrich their lives. It is Berger and Luckmann's view that the behaviours most likely to be reciprocally typified are those relevant to the castaways' common situation, such as eating and searching the horizon for passing ships, those that they bring to the social situation from their presocial lives, and activities relating to 'labour, sexuality and territoriality' – which sound like uncommonly biological points for sociologists to make, but bring us to the next development in the story.

A product of the reciprocal typification of habitualized sexual activity by Robinson Crusoe and Girl Friday is a bouncing baby girl. Following the thinking of the great founding fathers of their discipline, the sociologists Georg Simmel of Germany and Emile Durkheim of France, Berger and Luckmann point out that the change from a dyad to a triad marks a significant change in the quality of the group. What had been the rather *ad hoc* and informal institutions of Robinson Crusoe and Girl Friday are transformed into historical institutions with more crystallized objectivity, because now the institutions transcend any one person. The bouncing baby girl is soon followed by a baby boy, and when others are cast up on the island it isn't too long before Robinson Crusoe and Girl Friday are grandparents. Now the people being inducted into the island institutions are told 'This is how it is done', which embraces a kind of disembodied we-intentionality – especially when Robinson Crusoe and Girl Friday eventually die, as in time do their children, but the institutions live on.

This sequence from I-intentionality to we-intentionality and then to disembodied we-intentionality is the core of Berger and Luckmann's analysis. When Robinson Crusoe and Girl Friday were alone the routines they established had a certain flexibility about them. These were their own routines, which they could examine and change, even abolish. They remembered why they did certain things because of their shared biography, and so 'the world thus shaped appears fully transparent to them'. But then come the children, and as the parents transmit their routines to their offspring, the 'objectivity of the institutional world "thickens" and "hardens", not only for the children, but (by a mirror effect) for the parents as well'. The social world confronts the children as a given reality, as a set of external, coercive facts. This is how the social world is born, and this is how it is maintained. In Berger and Luckmann's words: 'An institutional world . . . is experienced as an objective reality. It has a history that antedates the individual's birth and is not accessible to his biographical recollection. It was there before he was born, and it will be there after his death. This history itself, as the tradition of the existing institutions, has the character of objectivity.' And each member of a social group locates themselves within that objective, historical world. The social world is there and cannot be ignored or wished away. Indeed, because social institutions exist as a reality external to the individual, they cannot be understood by introspection. 'He must "go out" and learn about them, just as he must learn about nature.'

Again we have the mystery of social reality being maintained down the generations through sheer force of shared belief, whatever the origins of that belief and how it is maintained. Legitimating formulae, Berger and Luckmann argue, are one of the ways such beliefs are maintained. 'We do things this way when we are hunting because women are better at it and so have always been our chief hunters,' say our islanders, and 'The men cook the food and do the dancing because these are activities that demand creativity and men are creative and women are not.' Well, legitimating formulae may go some way in accounting for continuity of institutional life, but humans are clever creatures and some will see through the frailties of the legitimation. They present a problem of compliance and this must be dealt with by sanctions. The institutions claim an authority of objectivity over the individuals, whatever their subjective interpretations of the world they live in. Children are taught what is right and wrong and how to behave, and adults are kept in line by a variety of means that range from mild disapproval to ejection from the group, be it by banishment or death. And all this is possible at a high level of abstraction because language provides the vehicle whereby institutional coherence is presented, justified and maintained. Language becomes the great means of objectivation of the social. The social world becomes a world of maxims, morals and narrative myths which reinforce and support that world, and different institutions interleave to strengthen further social reality. Hunting is tied to weapons and food preparation and consumption; weapons, in turn, are linked with the defence of the group against outsiders, whereas food is associated with fun and dance. Specific roles are created within each institution, and these, in turn, may come to symbolize specific institutions. The Queen is a great hunter and warrior, hence, as head of the hunting and defending forces, comes to represent them. No one knows any longer why the Queen rather than the King holds this symbolic position, and few question the superior courage and physical skills of women even though they seem not to throw objects with the accuracy of men (which hundreds of years later scientists will show with empirical certainty to be the case) and may be a little slower on their feet. The Queen's courage stems from her literal motherhood, say the myth-makers. No one remembers Girl Friday's unusual prowess – it is possible no one remembers there ever *was* a Girl Friday. Indeed, roles and the institutions they represent can and do become reified; that is, they are given meaning that gives them an ontological status independent of human agency. Hunting becomes associated with universal laws of subjugation, and those who hunt, especially the Queen, become impelled to act as they do 'because of their position'. Reification means that eventually 'man is capable of forgetting his own authorship of the human world', in Berger and Luckmann's own words.

As with Searle, so with Berger and Luckmann. The preceding paragraphs are but a small part of what has become a classic text on the sociology of knowledge. The two authors go on to analyse four levels of objectivation, which higher levels they term legitimations, the uppermost of which are symbolic universes that 'put everything in its right place' and in terms of which we try to understand the meanings of our lives. The second half of their work deals with the process of socialization by which children come to internalize the social world, about which more will be said in a later section of this chapter. This is the province of developmental psychology rather than sociology, but it is worth noting the two stages Berger and Luckmann put forward. The first, primary socialization, covers the period during which every child takes into itself the main features of social life and social reality. It has long been held by psychologists that a crucially formative relationship, known as attachment, exists between an infant and its principal caregivers. This emotionally charged relationship, or set of relationships, is the psychological crucible within which a child acquires its first subjective understanding of itself within the context of an embracing social reality. It is what gives socialization its emotional kick-start. Berger and Luckmann's second stage, secondary socialization, involves the internalization of specific institutionally based subworlds of roles, responsibilities and divisions of labour within specific social institutions, such as the family, friends, school, neighbourhood communities and the workplace, and the individual's relationships with these. At one point in their analysis the authors make an observation that recalls the metaphysical giddiness Searle describes as coming with a good hard stare at social reality. It is worth quoting in full: 'Primary socialization thus accomplishes what (in hindsight, of course) may be seen as the most important confidence trick that society plays on the individual – to make appear as necessity what is in fact a bundle of contingencies, and thus to make meaningful the accident of his birth.'

The 'confidence trick' that society plays is to present social reality to the child with much the same force as physical reality. Yet unlike natural objects such as mountains and chairs and people, which rest on a solid causal base within physical reality, social reality is poised on a seemingly unsubstantial and fragile base of what others agree is done or not done, is good or bad, and even exists or does not exist. Berger and Luckmann's confidence trick returns us to that mysterious, almost magical quality that culture has of bringing into existence a reality that has massive consequences for all of us yet seems to be based on almost nothing. Hence the assertion made at the start of this chapter: culture is imagination made real. Since human imagination is almost boundless, so too is culture.

Social representations

Implicit in the work of Searle and of Berger and Luckmann is another social-science thread that must be woven into our tapestry, which is that of social representations. These are representations that are shared by two or more people, and usually by a significant proportion of any social group. Social constructions are, in fact, social representations, albeit special ones because they bring into existence what does not exist outside social reality, whereas social representations can apply to any feature of the physical world of Searle's brute facts. The notion of shared representations first gained widespread currency following publication of Emile Durkheim's *Sociology and Philosophy* in 1898. Durkheim drew a sharp distinction between individual representations and what he called collective representations, a distinction roughly similar to that between, on the one hand, Wundt's experimental psychology and Dilthey's (and others') *Naturwissenschaften*, and, on the other hand, Wundt's social psychology (*Volkerpsychologie*) and Dilthey's sciences of mind, which were introduced in chapter 1. While the fit is not exact, it does give us some sense of what Durkheim was driving at. Collective representations for him were a form of social knowledge that could not be reduced to individual representations, hence could not be studied in the laboratory the way individual representations could be. They are a property of the social world, and that is where you find them, not in the psychological processes and mechanisms of each of us.

Again there is that odd and erroneous Geertzian feel that the social world is somehow materially shut off from the psychology and biology of the individual. While Durkheim was correct in that it certainly takes two to tango, nonetheless, while the we-intentionality of dancers cannot be reduced to the individual I-intentionality of the separate partners, it must be explained, and will one day be explained, in terms of the psychological characteristics and states of the people doing the dancing. Be that error as it may, for Durkheim collective representations are not just different from individual representations, they dominate human affairs and make us what we are.

Collective representations found their place in 20th-century social psychology as social representations, and in the last several decades one of the most influential proponents of social-representation theory has been the French social psychologist Serge Moscovici. It is not clear, at least to this observer, exactly how social representations are different from collective representations, but since the former term is dominant in the social psychological literature, it is social representations that we will be talking about. AIDS provides us with a good example of social representations. Roughly speaking, there are two clusters of social representations that refer to this disease. One is formed round the dominant scientific view that AIDS is caused by a virus that slowly destroys the immune system and makes potentially deadly illnesses out of infections to which healthy people would not be susceptible. The other is centred on the belief that AIDS is a disease which is a punishment for those who are guilty of unnatural and/or permissive sexual behaviour. Clearly these are not exclusive of one another, but people tend to identify with one or the other. Now, social representations are, to quote Moscovici, 'a network of ideas, metaphors and images, more or less loosely tied together, and therefore more mobile and fluid than theories'. They are not replicas or reflections of the world, they form the world. They become reality, and that reality may be complex. Thus, to the scientist, or the layperson who is not convinced of transcendent,

God-given morality, AIDS links with other social representations like pathogens, infections and possibilities of treatment and eradication of the causative agent. Those sharing the representations of punishment and excess also share representations relating to God, a righteous life and transgression.

This property of 'connectedness' is important to social-representation theorists for two reasons. The first is that it reveals the two processes that are claimed to determine social representations, namely 'anchoring' and 'objectification'. Anchoring is the process whereby strange facts and new ideas are brought into a familiar context and thus made normal and usual. When AIDS first made its appearance in Western countries in the 1980s it was a previously unheard of disease. Its virulence and the growing realization that, despite its initial restriction to specific groups of people, it related to a near universal behaviour, sex, made it especially disturbing. When the AIDS virus and its means of transmission were identified, the disease was incorporated into already existing and related representations of illness, either the possibilities of therapy, even immunization, or the alternatives of punishment for excess and the need to maintain moral codes of behaviour. Step by step the representations of AIDS changed and the unknown was transformed into the known and the accepted. Social psychologists view this as a form of defensive adaptation in the broadest sense. The very purpose of representations, Moscovici tells us, is 'to make something unfamiliar, or unfamiliarity itself, familiar'. But it is more than that. Anchoring also draws the individual into a particular social reality, and it is in childhood that the process of identification with a specific social group takes place. Objectification is a process whereby the abstract is translated into concrete images, the act of translation altering the representation and often associating it with specific objects in the world, as in totemization and other forms of symbolization and ritualization. The abstract is cashed out into the objective world of everyday reality and behaviour.

Nobody will be overwhelmed by the detail on offer for these putative processes. Moscovici confesses that a hundred years on we are little advanced from Durkheim, who complained that 'as to the laws of collective thought, they are totally unknown. Social psychology, whose task it was to define them, is nothing but a word describing all kinds of varied, vague generalizations with no definite object as focus.' However, this brings us to the second reason why the property of connectedness is important to social-representation theorists. As a consequence of anchoring and objectification we can begin to explain the seemingly inexplicable: how it is that our species is, as Moscovici puts it, 'the only animal gifted with reason [that] has proved to be unreasonable'. What he is referring to is the tendency we humans have of believing the absurd. Durkheim, writing in 1912 on religion, asked how could a 'hollow phantasmagoria have been able to mould human consciousness so powerfully and so lastingly?', and 'if the people themselves created those systems of mistaken ideas, and at the same time were duped by them, how could this amazing dupery have perpetuated itself through the whole course of history?' An objective 21st-century world view suffused with scientific knowledge rejects the notion that the world was created by a divine being, or beings, who now watches, or watch, over us; or that we have all lived previous lives and after our impending deaths will live reincarnated in another form; or that we each have souls that survive the death of our bodies. Yet religious beliefs live on, and in some parts of the world are expanding and deepening. So too are other forms of groundless belief. We are a constant witness to ridiculous views that range from the presumably harmless belief in alien abduction shared by an alarming number of Americans to deeply harmful racist conceptions according to which people passionately believe that they, and their social or ethnic group, are superior along a host of dimensions to others, and that their superiority sanctions violence and murder. How can it be that thousands of millions of people believe in the divine, or in the inferiority of others, when such views simply cannot be supported by either the evidence of their senses or by science?

Well, the reason why a conference of bishops declared AIDS a punishment from God is that they were anchoring and objectifying a new disease in terms of their dominant and familiar representations. This may seem trite and question begging. Claiming that people hold the beliefs they do because these were enculturated into them as children, and that enculturation is what is partly captured by anchoring and objectification, sounds like a banality of folk psychology and doesn't tell us why modern Britons no longer believe that getting up in the morning and painting themselves blue is the right thing to be doing but do believe that the monarchy is a good institution and that the existence of God should at the very least be taken as a serious proposition. This isn't quite the case. Social-representation theorists are not so undiscriminating as not to address the telling apart of the wood from the trees. They make a specific point, which, perhaps, is not so banal. It is this.

Some habits and beliefs do die and others persist. This must be explained. For example, some evolutionary psychologists argue that humans have a predisposition to certain sorts of beliefs and acts in the face of an incomprehensible existence but that painting oneself blue each morning is not one of them. Hence the persistence of religion but not of specific body decoration. Other evolutionists argue that religion is a by-product of cognitive modules that evolved for quite other reasons. Social scientists without any commitment to evolution claim that religion is associated with the human tendency to form particular kinds of social structures; others claim religion is just another creation myth born of the human need to locate ourselves within some comprehensible framework. Well, social-representation theorists are quite clear where they stand on these questions. Evolution is an irrelevance and society is not simply a collective of individuals no matter what distinctive social structure is formed, or what collective good social living is for. 'Certainly, power and interests exist, but to be recognized as such in society there must be representation or values which give them meaning,' according to Moscovici. Simply living together and sharing actions and knowledge are not enough. 'It is when knowledge and technique are changed into beliefs' with their accompanying symbols and rituals that what is just a collection of individuals is transformed into a social group, a culture, which is 'bound by a common passion and which is transmitted from one generation to the next'. Here again, if in slightly different form, is that strange quality, and once more it seems to be bound up with we-intentionality. In unspecified ways, social-representation theorists, or some of them, go beyond the we-intentionality of Searle and Berger and Luckmann as the source of the 'added force', the mystery, of social reality. It isn't just sharing, it seems to be sharing certain things in certain ways, but we are not told precisely what things beyond hints as to the power of normative representations, nor in what ways, though ritual is hinted at. In any event, social representations are closely bound together; some are representations of representations, and as such are only indirectly related to what is out there in the world. It is this tight clustering and this indirectness of representations that make cultures so impenetrable to outsiders who can't see the connections.

The Oxford philosopher Rom Harré raises some important points about social representations. One of these is best understood in the context of a study by the French social scientist Denise Jodelet, which concerns the social representation of madness derived from a study of a small town and its surrounding villages in France, where, over a period of several generations, the inmates of a nearby asylum were lodged in the homes of locals. Taking on this caring role generated significant income for the hosts, and the patients were thus an important economic feature of the commune, but they were nonetheless known to be mentally ill, and hence were thought of within a specific context of illness and the need for 'normal' people to keep a certain distance from them. These conflicting 'facts' about the patients determined both how they were understood and thought of by the people of the commune and also how they were treated, where they slept and ate, and so on. Thus Jodelet could study the social representations of madness as they had developed over some years and how they determined people's behaviour. She recorded that these representations were at once 'structured fields' of values, beliefs and images on the one hand, and 'structuring nuclei' that orchestrate the meanings of acts and observations on the other. Jodelet's conception of structure and structuring is very similar to Berger and Luckmann's dialectic of social reality. Representations are at once caused and causal, but quite how causation sits with regard to representations needs, Harré argues, to be sorted out. He also points out that representations are systems of signs and that there must be rules or conventions that govern their proper use. If we are to have anything approximating to a real science of social representations, these rules or conventions need to be established. Like Searle, Harré is inclined to look for them in the philosophy of speech acts, though that is not a direction in which we will go here. The point that Harré is making is that we need to move from a merely descriptive account of social representations to one that entails causation. Then we will have a science of social representations.

Social representations embrace all of social reality and the social world. Social constructions and higher-order knowledge structures like schemata and scripts are specific subsets of social representations. So, if we are to understand those most rarified of representations, social constructions, we have to do so by way of understanding social representations at large. Now, while many social-representation theorists are wary of what they call 'damaging cognitive reductionism', understanding social representations in terms of the cognitive processes and mechanisms that generate them is not reductive in the least, provided one recognizes that the nature and dynamics of such social entities can legitimately be studied in their own right. Jodelet's account of how 'the lunatics' came to be thought of in the French commune stands on its own terms. Nevertheless, links to the generating cognitive processes and mechanisms must exist, and must become known. Social-representation theorists give us absolutely nothing of mechanism and not much by way of processes. Anchoring and objectification are weak tea indeed. However, Moscovici and others do give a sense of the tight and complex interleaving of social representations (which nicely matches Searle's iteration of the basic structure of social reality 'X counts as Y in C'), and of the need to climb into the detail of these representations because that is the only way to understand a culture, along with some hints as to why some representations have greater potency than others. As a theory their approach is broadly descriptive and characterized by a massive claim on the potency of we-intentionality (a phrase they do not use and may not know) without casting any light whatever on where this comes from. They take it so much for granted that they just do not see any mystery or confidence trick, and certainly never seem subject to any feelings of giddiness when contemplating social representations. So if one is looking for an understanding of the source of we-intentionality within a causal framework of processes and mechanisms with a view to abolishing that giddiness, one must look elsewhere. And that elsewhere is the ideas of Vygotsky and other cultural psychologists.

Cultural psychology

It would not be accurate to describe cultural psychology as the science of enculturation alone. Enculturation, the developing into and becoming part of a culture, is, though, an absolutely central matter. How people, at the start of life, enter into a culture determines the future both of their individual lives and of that culture. So it is to enculturation that we shall confine ourselves in general, and to some of the ideas of Vygotsky specifically.

Lev Semenovich Vygotsky was born in 1896. A precociously and prodigiously talented child of Jewish parents who grew into a person of polymathic brilliance, he was lucky to be accepted into university given Tzarist Russia's overt institutionalized anti-Semitism, which restricted the numbers of Jews who could attend. He became a school teacher, but in 1924, after an astonishing presentation at a conference, he was invited to join a prestigious psychological research centre in Moscow. He died just ten years later. Although his work was ostensibly an attempt to create a Marxist psychology, it was clear that his ideas came from many sources, including, perhaps especially, literature and semiotic theory, and his work fell into political disfavour. However, his influence remained strong among Soviet psychologists, in part through the continuing studies of his two principal collaborators, A. R. Luria and A. N. Leontiev. With the intervention of the Second World War and the Cold War that followed, it took decades for Vygotsky's theory to gain a hold in Western psychology. For reasons that will become clear shortly, it was mainly in the psychology of education that he became well known, especially through being championed by the likes of the American developmentalist Jerome Bruner. It will also become clear that in contrast to the social-representation theorists, who have had little to offer by way of a theory of just how social reality is caused, Vygotsky and his followers provided relatively rich pickings in this regard because what he offered was nothing less than the beginnings of a theory of enculturation couched in terms of specific processes.

I want to focus on an exquisite conceptual creation of his, the zone of proximal development, as a means of exploring the possible origins of the transition from I-intentionality to we-intentionality. To get to the zone of proximal development, however, we must first pass briefly through Vygotsky's general conception. What follows, therefore, is a thumbnail sketch of his ideas, albeit an outrageously condensed one, for his original writings are not easy reading and are notoriously hard to simplify, not least because of the holistic nature of his approach, in which everything connects to everything else.

Vygotsky's theory was of massive scope, truly a general theory of the human mind. One other prefatory comment is necessary. While the theory has all the appearance of a general-process approach to cognition, it is not at odds with a domain-specific view of human intelligence, in which the cognitive mind is conceived as a set of innate organs each specialized to operate in a constrained way on specific inputs (see chapters 2 and 5). The computational outputs of such modules must interact at some point or points to result in coherent cognitive functioning. How exactly this happens must be a consequence of the architecture of human cognition, about which we presently know very little indeed. If Vygotsky's theory can be characterized by a single phrase, that phrase is 'tool mediated coherence of social cognition'. Not only does coherence not exclude modularity, it positively demands it. The transformation of different domains of input by computationally specialized neural networks has to proceed to a point where different inputs can be meaningfully related to one another and form the basis for reasoning, storage of memory, retrieval of already stored memory, decision-making, planning for action and executing plans. Vygotsky's theory concerns what happens after different domains of information have been transformed into neural codes that can be related to each other. This is not, of course, how Vygotsky would have put the argument 70 or 80 years ago. He did, though, realize that, for example, 'thought does not immediately coincide with verbal expression', indeed that 'thought is inexpressible'.

Vygotsky well understood that the human mind must be located within an evolutionary framework. Tool use by some other apes was well known to him through the work of Köhler and others, but he believed that humans use tools in unique ways, as a means to shape and change the world within a framework of exploratory and expanding use which ends in precise planning and execution. Chimpanzees can use two stones to crack open a nut, but cannot fashion a hammer from different materials to drive a nail through two objects in order to hold them together and thus provide part of the frame of a structure for dwelling in that will comprise scores, even hundreds, of other units of construction. The evolution of such tool use, which goes way beyond anything any living non-human ape is capable of, was, in Vygotsky's view, a, perhaps *the*, crucial event in human phylogenesis. He also understood very clearly that this phylogenetic characteristic becomes manifest in, and is a crucial part of, the ontogeny, the individual development, of the mind of every child. And crucial to the ontogeny of the mind are its interactions with other minds, minds that have themselves been shaped by their social reality, their culture and the history of their culture. Thus four sources of influence, or 'levels of history' as Vygotsky put it, the phylogenetic, the ontogenetic, the cultural and the historical, all converge as forces in the creation of the mind of each and every person, and they do so by way of the minds of others. 'The whole history of the child's mental development teaches us that from the first days, his adaptation to the environment is achieved by social means through the people around him,' he wrote. 'The path from the thing to the child and from the child to the thing lies through another person', the 'thing' being an action, a meaning, an object, a ritual, a solution to a problem or anything else that can be learned, including language. Unlike Jean Piaget, another of the great founding fathers of cognitive psychology, for whom cognitive development is a rather lonely journey through the world, Vygotsky saw the growth of the mind as a process of constant social transaction.

Vygotsky did not, unlike most other psychologists at that time who knew that apes could use tools but have no language, distinguish on that basis between tools and language. To him, language, like other symbols, is a tool. Indeed, it is 'the tool of tools', because it does not merely stand within Searle's structure of 'X counts as Y in C', but it also plays a crucial role in mediation. Mediation is a central process notion within the Vygotskian scheme and is perhaps most easy to understand in the context of speech. For example, if during play a child is seen by her mother to be doing something potentially harmful and is told 'No, Jane, don't do that', Jane may subsequently say to herself, out loud, 'No', as a means of regulating her own behaviour.

The mother's speech mediates Jane's egocentric speech, her speaking to herself, which further mediates Jane's regulation of her own behaviour. Egocentric speech mediates the internalization of speech, and inner speech mediates further forms of thought and problem-solving. Unlike tool use by non-human apes, tool use by humans, including the use of language, is a process of bridging and building upon other processes. In fact all behaviour is not just directly applied to some particular task, it also assumes the auxiliary role of mediating in that task, and may then mediate further psychological processes and further behaviour. This power of mediation extends into the inner functioning of the developing child's mind. Lower concepts mediate higher concepts, and these higher concepts then mediate the transformation of those lower concepts. The child can use the high ground of a new idea to transform her understanding of the lower concept that got her there. For example, learning some algebra and geometry alters the way simple arithmetic functions like addition and multiplication are understood and performed. One of the original features of Vygotsky's theory is that mediation results in novel processes and new states. Mediation results in constant change in the child's cognitive states. Having reached some higher concept, which is a new state, the lower concept, the old state, is changed as well. This conception of a dynamic and constantly changing state of cognition contrasts most markedly with the work of Vygotsky's compatriot, Ivan Pavlov, and of the Western associationists so influenced by Pavlov, who believed that significant amounts of human knowledge come about by a simple accretion of connections between representations of the world. This was not how Vygotsky believed cognitive development occurs. He could not specify mechanisms the way Pavlov did, but the processes of change in cognitive structures were quite different from just a simple linking of things one to another. 'Children solve practical tasks with the help of their speech, as well as with their eyes and hands,' wrote Vygotsky. 'This unity of perception, speech and action, which ultimately produces internalization of the visual field, constitutes the central subject matter for any analysis of the origins of uniquely human forms of behaviour.'

Because social interaction was for Vygotsky the absolute essence of cognitive development, history enters the story of enculturation through the transmission of tools, including of course language, across generations. Each 'parent' or older peer impinges upon the psychological development and enculturation of the child by demonstrating tool use, pre-eminently the use of language and other symbols. Thus human cognition is intensely social in its origins, with others mediating between tools and the child, and the tools mediating in the cognitive development of the child. Culture is an accumulating pool of artefacts, of the things humans construct, including all social representations. Michael Cole, the American cultural psychologist, nicely defines culture as 'the mediation of action through artifacts', a point to which we will return.

There is much, much more to Vygotsky than the above few paragraphs cover, including his writings on play ('the main channel of cultural development'), consciousness (which he did not understand, but then no one does to this day), education, numerous experiments he and his colleagues carried out, and even the compilation of ethnographies in the search for differences in cognitive styles among people undergoing rapid cultural change during the time of collectivization in the Soviet Union of the 1930s. The main point to make is that fractionated understanding of the mind was not what he was after. Reaching into literary theory he invoked a complex inversion of relationships.

Understanding the words of others also requires understanding their thoughts. And even this is incomplete without understanding their motives or why they expressed their thoughts. In precisely this sense we complete the psychological analysis of any expression only when we reveal the most secret plane of verbal thinking – its motivation.

Thus, for Vygotsky, every psychological function appears on two planes. There is the social, interpsychological plane, and there is the intrapsychological plane within the mind of the developing child. Social interactions underlie and mediate all psychological events and states, which then manifest themselves in new interpsychological interactions. Armed with these broadly based ideas we now have enough of a flavour of Vygotsky's theory to move on to his zone of proximal development.

Vygotsky illustrated this concept by invoking two children, both of whom have been assessed as having the mental age of an eight-year-old.

We do not stop with this. Rather, we attempt to determine how each of these children will solve tasks that were meant for older children. We assist each child through demonstration, leading questions, and by introducing the initial elements of the task's solution. With this help or collaboration from the adult, one of these children solves problems characteristic of a twelve year old, while the other solves problems only at a level typical of a nine year old. This difference between . . . the child's actual level of development and the level of performance that he achieves in collaboration with the adult, defines the zone of proximal development.

Thus the zones of proximal development for his two hypothetical children are different. What is important is not so much where you *are* cognitively but where you are *going to get to*, and 'the zone of proximal development has more significance for the dynamics of intellectual development and for the success of instruction than does the actual level of development'. A few pages later, in the same essay, he wrote:

What lies in the zone of proximal development at one stage is realized and moves to the level of actual development at a second. In other words, what the child is able to do in collaboration today he will be able to do independently tomorrow.

So what is plain is that the zone of proximal development is not some fixed place in cognitive development or cognitive space but is constantly moving like a wave of cognitive function that sits just in front of where any child's cognitive abilities are. But, and this is the really important thing, the zone of proximal development is not a wave that is being pushed by the child so much as a wave that is being pulled by the other person who enters into it, an adult or superior peer.

Consider the definition of the zone of proximal development by Vygotsky that is more usually given:

The distance between the actual developmental level as determined by independent problem solving and the level of potential development as determined through problem solving under adult guidance or in collaboration with more capable peers.

In effect, this places within a psychological context the Socratic method that Plato illustrated in his dialogues nearly two-and-a-half thousand years before Vygotsky put pen to paper. What Socrates did was build upon the knowledge someone had of a matter, such as the sources of human nature or of geometry, by asking probing questions that helped that person to establish a greater degree of knowledge and understanding. Indeed, the metaphor of the Socratic method gives the child's contribution within the zone of proximal development a less passive appearance than some of Vygotsky's writings would suggest.

In addition to its pedagogical role, however, Vygotsky made a very specific assertion about the zone of proximal development: 'Human learning presupposes a specific social nature and a process by which children grow into the intellectual life of those around them.' This tells us that the zone of proximal development is an intensely social place. It isn't just learning and education we are talking about here. It is learning and education within a social context where the presence of others in relation to the child is absolutely essential and central. It is where all social representations are born and propagated. Bruner has developed further the notion of the zone of proximal development around what seems to be a puzzling feature in Vygotsky's formulation, which is that the child's conscious control of an act, a word or a meaning comes only after that act, word or meaning has been mastered within the zone. Bruner suggests that this happens through a process of 'scaffolding'. What a good tutor (the adult or superior peer who is leading the child through the zone) does is provide a vicarious consciousness for the child until the child masters the action. The tutor literally thinks for the learner, functioning like a scaffold to tie the tutor's conscious control to the child's developing internalization of the knowledge of the task and its conversion into a tool for the child's own conscious control. This leaves a great deal to be explained, but it provides a vivid image of what enculturation might be about.

Think how primary caregivers, usually parents, interact with babies almost from the moment they are born. An adult faces their baby and, usually taking their cue from it, initiates an interaction by moving their head, adopting an exaggerated facial expression, holding the baby's hands and moving them, and vocalizing in a variety of ways, including, of course, using language. This elicits a response from the baby, and that response is responded to with further activity from the adult, which in turn brings a further response from the infant, and so it moves back and forth. Later, when the infant is some months older and beginning to babble in a more controlled way or beginning to utter single words or two-word sentences, the adult initiates a more concentrated vocal interaction, often, though not universally, using a form of language sometimes referred to as motherese. Motherese is a form of speech that adults in many cultures adopt, seemingly unconsciously, when talking to babies. The speech rate is slowed, the pitch is exaggerated, and the grammar is simplified as the adult addresses things that are impinging upon the infant in the here and now. (It is interesting to note that children as young as five or six years old will also adopt a form of motherese when talking to a younger child who is at a protolanguage level of linguistic skill.) The adult picks up on something the baby has said or babbled and responds with this exaggerated and simplified form of language, and again the interaction goes to and fro. Thus the adult, without any conscious planning to do so, tailors their language to the infant's ability.

Bruner describes this as language learning of a 'formated, pragmatically paced kind ... that is ... governed by a rule of "voluntary handover and willing receipt"'. It so impresses him as to lead him to conclude that there is a 'Language Acquisition Support System (LASS) that is at least partly innate'. LASS, Bruner suggests, is a counterpart to LAD, the innate Language Acquisition Device that Chomsky, at an early point in the formulation of generative-grammar theory, suggested might account for language learning. In the domain of language and learning, LASS and LAD operate within the zone of proximal development, with LASS providing the scaffolding on which the language learning of the child develops. Later still in the child's life, the parent may look at a picture book with it, pointing to illustrations and saying things like, 'Look at that! It's an elephant. What is it? An elephant, that's right.' More of this constant to-ing and fro-ing – back and forth goes the interaction, with the tutoring adult, or superior peer, guiding the child through the zone of proximal development.

Michael Cole points to parallel cases of the zone of proximal development in cultures different from our own. In African Tale society education is 'regarded as a joint enterprise in which parents are as eager to lead as children to follow'. The Kpelle children of Liberia, in age-graded groups, are led by adults or older children through verbal games of storytelling and riddling in the course of which salient features of Kpelle culture are revealed. The Zinecantecan weavers of south Mexico enculturate their children into weaving skills in a specific sequence, with close initial supervision and teaching slowly giving way to a more relaxed and distant overview of the children's work. The same general structure applies to the teaching and learning of tailoring skills in Liberia. In each case a zone of proximal development is established and in a near errorless learning procedure the adults or more competent peers gradually lead the children from where their cognitive skills and activities are to some further point of advancement.

The notion of the zone of proximal development has obvious importance within educational settings, both formal and informal. However, it might also provide us with the means for beginning to understand the mystery – Berger and Luckmann's confidence trick – of how a child, acquiring we-intentionality and abstracted or disembodied we-intentionality, grows with a natural acceptance into the seemingly fragile, though actually incredibly powerful, social reality that characterizes its culture and will dominate its life. Within the zone, turn-taking rules, whatever is being learned. First there is egocentrism; but soon I or me gives way to you and me, and to me and you. You and me gives way to us, and us mediates a change in the conception of you and me. Us is internalized as we, and as the theory of mind mechanism develops (as described in chapter 5), we move over the first three or four years of life through the dimension of I-intentionality to an understanding of you-intentionality and on to we-intentionality. The development of we-intentionality has been little documented and studied, in contrast to the flood of recent work on you-intentionality (Theory of Mind), so there is no basis at present for knowing just how it relates to the latter. It probably does not occur as a simple linear sequence, with we-intentionality following straight on from the acquisition of you-intentionality. The fact is, we just do not yet know. And within the zone plenty of things are happening at the same time, and we do not yet know just how each is dependent upon others. Language and other non-linguistic skills are being learned, as is an increasing command of number and an understanding of physical causation; the gradual building of higher-order knowledge structures is occurring and new physical skills are being acquired, and those already mastered are being further honed into areas of expertise. In many cases the learning, often in the context of play, is characterized by that ubiquitous turn-taking. What the zone of proximal development might provide the child with is a generalized conceptual arena within which, through turn-taking, within the common framework of give and take, a more generalized capacity develops which allows it to enter into collective intentionality, as defined by Searle. From this the child grows into the social reality of its culture and comes to accept the culture's customs and conventions in line with the collective agreement of what these are, without staring at this agreement too hard and wondering whether things really can or should be the way they are. Maybe the magical power of social reality is born within Vygotsky's zone of proximal development. This is just a hypothesis, and a fairly crude one at that. But it is an example of the kind of theory structure we need, one that moves us from mechanisms of mind to the process of enculturation, which we are going to have to understand better if we are ever to have a science of culture that stretches from the natural sciences to the myriad forms that social reality takes.

A final word on Vygotsky's theory and its bridging role. Cole's definition of culture as 'the mediation of action through artifacts', which embraces language and other symbols and social constructions of cultures, gives us, as Cole notes, a link to the 'thick' description of the symbolism of Geertz, and also to Jodelet's social representations of madness. Cole's link is Searle's bridge between physics and the social sciences, and that is a bridge that must be made up of psychological processes and mechanisms. Vygotsky's psychology, in its original form and in the way it is being currently shaped by cultural psychologists, is almost certainly wrong in detail and probably misses hugely important things that have yet to be discovered. But its general shape and formulation will be what we will come to understand as the basic construction of that bridge, a construction straight out of the natural sciences, those non-optional forms that Searle lays down for understanding the world, by which we will reach a richer understanding of culture and its strange qualities.

A tentative conclusion

Social reality does exist, contrary to the views of some biologists who think it all an invention of muddle-headed social scientists. After all, these hard-nosed sceptics have work contracts with their universities or research institutes, citizenship of a country and money in their pockets – all instances of the social reality that rules their lives. These are products of collective intentionality that serve functions whose causal force resides in agreement between people and not in their physical embodiment as written contracts, passports or bank notes, and which are bound together by constitutive rules. For Searle, we-intentionality is at the heart of social reality, and Berger and Luckmann provide a nice demonstration of the dimension that begins with I-intentionality and ends with disembodied or abstracted we-intentionality. But where can we begin to place the source of we-intentionality, and what are its mechanisms? For Vygotsky and his followers, cognitive development *is* enculturation. The individual's own activity, whether physical or mental, is central to enculturation, which, most vitally, is something that happens in a social context. And Vygotsky gives us a 'place' in cognitive development for this, the zone of proximal development. But is this at all plausible? Is this a bridge that will collapse the moment anyone steps onto it? Well, perhaps. But it is not quite so fragile a structure as it may first appear, if only because it ties in with the biology of previous chapters and begins to answer some of the criticisms directed at attempts to naturalize culture that have been made by social scientists and others sensitive to the complexity of the task.

The mantra of early chapters in this book is that culture is a particular and extraordinary manifestation of human intelligence. Vygotsky and his followers have long understood this, so the assertion is not original. Human intelligence, however, in an important sense, is not an exception, so has to be understood in the context of all forms of intelligence, which is an ancient set of adaptive traits that has evolved in response to certain rates of change in the world. But the importance of what is changing and the rates of change are relative to species' characteristics. That is why intelligence is always constrained, and that constraint must always begin, at least in part, in the genes that link every creature to its evolutionary past. This is not, remember, the reduction of intelligence to genetics, and the very existence of intelligence denies the possibility of genetic reduction of behaviour that is the product of intelligence. But it is a strong claim that intelligence in any species has been moulded by past evolutionary forces, even if the past is usually another country that we cannot reach. The principle of constrained intelligence remains. And that claim allows us to understand something of why the constraints on intelligence in any species take the form they do. We may not ever be able to provide a full account of the constraints precisely because we cannot with any certainty reach into the past, but the principle of constraint gives us some purchase for explaining the intelligence that we see now, *provided the particular form of intelligence that we see now in a species can be related to other species' characteristics in the present*. Hence the elegance of the study of differences in learning in closely related species of vole described in chapter 2.

What, then, of human intelligence? What is the equivalent in our own species of the demands for spatial learning in polygynous voles, and how does it relate to culture, that most extravagant feature of human intelligence? Again the answer is unoriginal. It is our sociality. This does not mean that our intelligence is limited to the social domain, because quite clearly it is not. Nor does it mean that social intelligence is limited to our species, which equally clearly it is not. But sometime in the last five million years our ancestors took an evolutionary turn within their social condition that led to our intelligence supporting culture as we see it now in so many different forms. Exactly what that turn was will always be a mystery, but it must, it simply must, have been connected with the advantages of group living, which led to the selection of certain features of intelligence as group-level adaptations. As we saw in chapter 6, group-level selection is a defensible and increasingly accepted idea. In this context, we-intentionality is a form of D. S. Wilson's social control, a psychological characteristic that reinforces the maintenance of our acting in concert with others and not behaving solely to further our own individual interests.

There is an even more striking link to be made with Boyd and Richerson's notion of a human genetic predisposition to choose cultural variants on the basis of frequency. Reading the social-science literature, especially that on social representations, gives a strong sense of the perceived importance of numbers, of how many people are doing or believing something. This is how Moscovici puts it: 'It is true that each person, in worshipping a plant or an animal, seems to be the victim of an illusion. But if everyone together recognizes their group in this way, then we are dealing with a social reality.' Durkheim made a similar point: 'It . . . subsume[s] the variable under the permanent and the individual under the social'; and elsewhere: 'Creating a whole world of ideals, through which the world of sensed realities seemed transfigured, would require a hyperexcitation of intellectual forces that is possible only in and through society.' 'Hyperexcitation of intellectual forces' is a curious phrase, but it resonates powerfully with the notion of social force discussed in chapter 5. If we are inclined to follow what the majority are doing or believing at least some of the time and to identify with that majority, the mysterious added value of culture might be accounted for by a simple predisposition of the kind modelled by Boyd and Richerson. Put another way, if culture is imagination made real, one of the forces that compels each of us to believe in and adhere to extraordinary imaginings is sheer weight of numbers of others believing in them. This does not mean we are all enslaved to the majority all of the time. It merely means that most of us are so predisposed some of the time. This predisposition is sufficient to begin to explain both the commonplace of enculturation as well as some of the appalling events in human history. If Boyd and Richerson are correct, this is an aspect of human psychology that needs to be set within, and balanced against, another feature of ours, which is to fission into subgroups, and to do so on almost any basis, be it a matter of allegiance to religion, ethnicity, political party or even a football team.

Another and obvious link to emphasize is that between enculturation in the zone of proximal development and the other mechanisms discussed in chapter 5. Of these, a Theory of Mind mechanism is of pivotal importance. If autism and Asperger's syndrome are indeed the results of impaired Theory of Mind, it is clear that the development of language qua language proceeds in the face of such impairment. But it is also clear that language in the full social sense of communicating thoughts and feelings is deficient, because those who suffer these illnesses are profoundly isolated socially. Whilst as yet little studied, there is an empirical programme awaiting researchers on just how autists are impaired in the process of enculturation, particularly in the degree to which their understanding of symbolism is deficient, as well as the extent to which they are or are not able to cope with higher-order forms of knowledge and enter into the social reality of social construc- tions, which owe their existence solely to collective agreement, to the ability of individuals to enter into the minds of others so freely and so powerfully as to form collective intentionality. After all, if we can come to understand that others have minds, we can surely come to understand almost anything, even the most bizarre of social construc- tions, such as a piece of paper having value.

Recent findings by neuroscientists, also mentioned in chapter 5, provide further intriguing links. The discovery of mirror neurons by Giacomo Rizzolatti and Vittorio Gallese of the University of Parma a few years ago just might turn out to be one of the bases for a neuro- science of culture. Mirror neurons, remember, are nerve cells in the prefrontal region of a monkey's brain that become active both when the animal is executing a specific movement and when it is observing others carrying out the same movement. The fit to action is quite circumscribed. These are neurons that do not fire simply at the sight of an object and/or a hand. The hand must be doing something specific to the object. So what was previously thought of as a part of the brain involved in planning actions turns out also to be a part of the brain that is sensitive to the actions of others. In 2000, Marco Iacoboni and colleagues, at the University of California, in Los Angeles, reported in the journal *Science* that a part of the human brain called Broca's area, which is known with certainty to be involved in speech production, becomes active when a person attempts to imitate the actions of others. (Brain-imaging techniques were used to monitor the activity in normal people.) The premotor cortex of monkeys is thought to be closely related to Broca's area in humans, and the possible link between hand movements and language is redolent of the claims of those who, like William Stokoe, believe that hand gesture was a precursor to spoken language. Most intriguing are functional-imaging studies that impli- cate the medial prefrontal cortex in thinking about both one's own mental states and the mental states of others. A study of people with brain damage, by Donald Stuss and colleagues at the Rotman Research Institute in Toronto, published in 2001, points to the right inferior medial prefrontal cortex as being involved both in simple perspective taking (understanding what others can know) and in detecting deception by others. The search for mechanisms is closing in. We are, however, a long way from knowing even at a crude level of brain location just how imitation, language and Theory of Mind are tied in with one another; but it is just possible that what is unfolding in the neuroscience literature is the explication of we-intentionality in neurological terms. What is badly needed is a neurology of social force, because compliance with the views of others, and a willingness to identify with and learn from others, seem to be two of the essential, if somewhat unsung, components of enculturation. Just which parts of an infant's brain are active, one wonders, as it is drawn through its zone of proximal development by a gesticulating, vocalizing adult? Functional-imaging studies and time will tell.

So much for some of the links that can be made between the social reality of the social psychologists and sociologists, psychological and neurological mechanisms, and evolutionary theory. It may be decades yet before we can say, with the same certainty that we say we know the internal structure of the atom, that we know 'social constructions are group-level adaptations that arise from the workings of x, y and z psychological mechanisms, which are present in m, n and o regions of the brain, which are linked with particular suites of genes located on chromosomes 2, 3, 7, 9 and 11'. But in an age when serious plans are being made to send people to Mars, and the history of the universe is known with increasing depth to within seconds of its origin, it would be unwise to bet against it happening.

But what of the criticisms that social scientists make of the naturalization of culture in general, and specifically the criticisms levelled against memetics listed at the end of chapter 4? There is a general answer to the question which breaks down into specific responses. The general answer is that in substantially taking on board the views of social scientists and philosophers on the centrality of social representations and social constructions, we keep the key elements of the social scientists' approach to culture intact. This means accepting that culture cannot be reduced to a homogenized set of entities and phenomena each of which is to be treated as equal to every other. A popular tune is not the same thing as a belief in God. If they are both memes, they are memes of such great difference that little is gained by giving them the same tag. One of them, the religious belief, is usually acquired just once in a lifetime during the process of primary enculturation. Mediated by language, reinforced by rituals and serving as a strong signal of group identity, it has all the characteristics of a powerful social representation. Popular tunes, on the other hand, have none of these characteristics – which is why they come and go. Whether there are features of cultures that are universal is an interesting question. Given the position adopted here, that what is common to all humans are the forms of intelligence that give rise to culture, hence that culture itself is universal, and given the commonality of certain problems to all human groups, such as the need to prepare food or to regulate relationships between the individuals that make up a social group, it should be no surprise if some features of all cultures look surprisingly similar. But such similarities cannot be taken as evidence of evolved cultural universals with deep biological significance. It is human nature which is universal, and sometimes so are the problems we face. Similarities between cultures are probably the result of convergent cultural evolution.

As to the atomization of culture and its decoupling from institutions, these may be legitimate criticisms of some biological approaches, but cannot be held against an approach that embraces the social reality of Searle, Berger and Luckmann and others. And if you take on board the cultural psychologists, history as one of the forces determining culture is right there in the zone of proximal development of each and every one of us.

Finally there is the central role of collective intentionality and the power of social force. Accept the importance of these, which these pages do, and another criticism of evolutionary approaches to human psychology and culture falls away. A charge often levelled against evolutionary psychologists is that the manifest diversity of humanity is ignored in the rush to establish evolved universals. Well, that is a charge that cannot stick if the processes and mechanisms invoked, and invoked as evolved universals, are the very things that cause human diversity. One tentative conclusion, then, is that this is not a bridge between the natural sciences and the social sciences that will collapse at the first critical footfall.

But why, some will ask, all that evolution? Why not, given acceptance that sometime in the last five million years our species evolved the capacity for first culture, and given acceptance that further evolution then brought us to something like contemporary culture around 10 to 12 thousand years ago, but given also that we simply cannot know with empirical certainty what the sequence was in the evolution of the vital mechanisms we now have, or what selection pressures drove their evolution – given these things, why not concern ourselves just with those mechanisms? The answer was given back in chapter 1, but the interposition of some 100,000 words makes it worth repeating. The theory of evolution is the central theorem of biology. Without it nothing makes complete sense, because it tells us about origins, design and function. An understanding of social constructions limited to their underlying psychological processes and mechanisms will take us quite a long way, but it will not give us a full answer. For that we need to understand why intelligence ever evolved at all and why it is always constrained in the way that it is. If culture is an expression of human intelligence constrained by the advantages of a group-level adaptation, and it surely can be nothing else, then we can only fully understand it within the broader conceptual framework of evolutionary theory. That theory will not give us the nitty-gritty understanding of the nuts and bolts. Only the proximate sciences of psychology and physiology will do that. But it will give us a grander view of a very grand thing indeed, which is what human culture is.

Suggested Readings

Berger, P., and Luckmann, T. (1966) *The Social Construction of Reality*. London, Penguin Books. (A classic sociological theory of social reality.)
Cole, M. (1996) *Cultural Psychology: A Once and Future Discipline*. Cambridge, Mass., Harvard University Press. (A readable survey of cultural psychology.)

Flick, U. (1998) *The Psychology of the Social*. Cambridge, Cambridge University Press. (An anthology of essays on social representations.)

Geertz, C. (2000) *Available Light: Anthropological Reflections on Philosophical Topics*. Princeton, NJ, Princeton University Press. (Essays on anthropological isolationism, among other things.)

Searle, J. R. (1995) *The Construction of Social Reality*. London, Allen Lane. (A philosophical analysis of social reality.)

Wertsch, J. V. (1985) *Vygotsky and the Social Formation of Mind*. Cambridge, Mass., Harvard University Press. (A summary and modern extension of Vygotsky's theory. Those wanting to read original Vygotsky will find some of his writings in six volumes published from 1987 to 1999 by Plenum and Kluwer, edited by R. W. Rieber and others.)

Index